CRAFT BEEF

CRAFT BEEF

A REVOLUTION OF SMALL FARMS AND BIG FLAVORS

JOE HEITZEBERG, ETHAN LOWRY, AND CAROLINE SAUNDERS

NEW DEGREE PRESS

CRAFT BEEF

A Revolution of Small Farms and Big Flavors

ISBN   978-1-64137-100-1   *Paperback*
ISBN   978-1-64137-101-8   *Ebook*

To farmers everywhere.

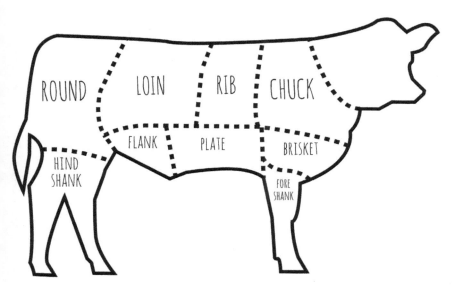

CONTENTS

INTRODUCTION 3

1. HAPPY COWS TASTE BETTER.. 15
2. SELECT—CHOICE—PRIME—*CRAFT*..39
3. THE OTHER 88 PERCENT 73
4. A SECOND CHANCE FOR SMALL FARMS 105
5. UNSUNG ENVIRONMENTAL HEROES 119
6. A FLIGHT OF BEEF..137
   ACKNOWLEDGMENTS..161
   APPENDICES...165
   NOTES 171

INTRODUCTION

———

For Ethan, whose wife and kids are vegetarians, this whole, big, beef *thing* started one evening after work when he was standing in front of the refrigerated aisle at the grocery store, squinting down at packages of steak. If he were going to bring meat home and prepare it under the watchful eyes of his non-meat-eating family, he thought, it had better be some seriously ethical, humanely raised stuff.

He scanned the brand names on the rows of cellophane-wrapped steaks—New York strips, ribeyes, and filet mignons—searching for a company that seemed to produce "good" beef (whatever that meant). Most of the brand names hinged on words like *farm* or *ranch*; some sported pastoral images of a grassy barnyard to match. The labels and marketing claims, from *organic* to *grass-fed* to *pasture* to *outside*, seemed meant to reassure, but their ability to soothe the shopper stuttered upon further inquiry. That's what Ethan found, anyway, when he consulted the meat cutter behind the deli counter:

Just how "outside" had these cows been?

Did they eat grass their whole lives, or just for a while?

Yeah, I see the marbling, but is it actually going to taste good?

The young man behind the counter blanched. He couldn't say what country the steaks Ethan was holding were from, much less what conditions the animals they came from had lived in, nor even how good the meat would taste.

Pulling out his iPhone, Ethan did a quick, hopeful Google search of the brand whose steaks he was still holding and found that the romantic name was a bit of a front: This beef came not from a farm but from a feedlot, just like all the other beef on the market. It wasn't even from the U.S. (Neither of those tidbits was apparently mandatory to disclose.)

Ethan went home steak-less that night.

Joe too was familiar with the frustrations caused by grocery-store steak. Since age 20, when his homestay family on a daikon radish farm in Hokkaido, Japan, roasted a Wagyu steer for his birthday, he had nursed a lifelong love of really, really good beef. In the intervening years, few steaks had lived up to that delectability of the farm-raised, umami-flavored, exotic-breed cow.

So when, a few years ago, a mutual friend of Ethan and Joe's—who worked with Joe at the time—walked into the office muttering about steak, Joe's ears perked up.

He asked, "What was that, Brendan?"

To which Brendan replied, bouncing on the balls of his feet, "I'm getting my cow on Friday!"

Joe didn't entirely know what to do with that announcement.

Brendan proceeded to tell Joe all about the annual cow he'd been "sharing" with a few neighbors each year for going on a decade. Brendan drove to the farm and spent 10 hours in his (now meat-scented) Prius to retrieve the beef. Far from the gastronomic mediocrity of most steak you find these days, Brendan assured Joe, this direct-from-the-farm cow was apparently, "Incredible. Absolutely incredible-tasting."

Joe was hooked. But when he asked if he could buy a steak or two from Brendan, the answer was a firm no. (Brendan's excited face reverted to a mask-like expression.) The resource—all 100 pounds of Brendan's share—was apparently far too precious for anything more than the pre-agreed divvying up among the neighbors.

A few nights later, still smarting from the loss of the apparently heavenly steaks, Joe was midway through a start-up brainstorming session with Ethan when he brought up Brendan's "cow-pooling" tradition.

A light bulb went off in Ethan's head:

"We should crowdfund a cow."

It was a relatively simple idea (or so we thought back then). Crowd Cow—as we christened our notion—borrowed the concept of cow-pooling, or sharing a cow among a few friends, moved the process over to the internet, and widened the pool to include up to 50 strangers. One cow would go "on the block" at a time; random people would lock in whatever cuts they wanted; and when the cow "tipped" (when all its cuts were spoken for, that is), all those steaks and roasts, and all the unusual parts too, would zoom off across the country to the "steak holders" who'd ordered them.

As we sketched out the idea, excitement building, we were stopped short by a shared realization: We were channeling a scene from *Portlandia*. You probably know the one.

The proverbially "Portland" diners, after bombarding the waitress with questions about the roast chicken's origins, politely insist that they halt the meal until they can visit the chicken farm themselves.

("Can you hold the table?")

The scene is hyperbole, a parody of the earnestness that runs through dietary decisions on America's crunchier coast. But like all good comedy, it also strikes a chord. The dogged line of inquiry after Colin the chicken's living conditions, and the motivations of the farmer who raised him, invokes our collective unease around the state of American food production—especially animal livestock. That hasn't gone away since the episode's 2011 air date. Even more than a few years ago, Americans today are hungry for meaningful connection to what they eat, more than product labels like "Organic" can provide.

So we drove forward with our idea but promised ourselves we'd keep things a little more lighthearted than the *Portlandia* characters had.

When we launched the very rudimentary website, the first cow sold within 24 hours, before we had even figured out whom to call about dry ice or whether what we were doing was legal under USDA regulation (don't worry, it was). We soon realized, as the second, then the tenth, then the hundredth cow sold, just how many people had the same instinct that it was time to go straight back to the farm for their steaks. Our idea had several perks:

1. Claiming a share of a cow on the internet from a farm whose story you could read about was kind of delightful.
2. You didn't actually have to get in the car and drive multiple hours to get great steak.
3. It was a tonic for our discomfort with America's notoriously opaque and problematic beef industry.

*   *

For Ethan and Joe—and later for Caroline, who joined Crowd Cow as our chief writer just as we expanded out of the Pacific Northwest and to the East Coast—the next two years became all about beef. Evenings, we pored over industry tomes on butchery cut plans. (Steak #1184d or roast 171g, anyone?!) Rainy days, we hunkered around farmers' dining room tables, talking economics and forage chains and soil organic carbon. Dry days, we tagged along (unhelpfully) on cattle drives in Pennsylvania or attended Kobe tastings in Japan.

After some time immersed as we were in the world of beef, there came a point when the three of us paused, looked around at each other, and realized we'd stumbled into something more significant and bigger picture than just great-tasting, transparently sourced beef. We'd walked smack-dab into a coalescing movement for which culinary America didn't yet have words. We called it "craft beef."

None of the steaks any of us had eaten in our pre-Crowd Cow lives had hinted that craft beef existed. Before we started this journey, we'd heard stories of the dismal goings-on at factory farms. But as we looked deeper, we learned that 50,000-cow-strong feedlots extend across the Plains in vast swaths of dirt and manure, impacting everything from groundwater[1] to antibiotic resistance[2] to atmospheric climate change,[3] causing illness in 30 percent of the cows that enter their gates,[4] disempowering small farmers,[5] and—perhaps most surprisingly—affecting the taste of steak. Because big beef is shooting for low cost to the consumer, good marbling, and high yield (super cows), they've headed down a path lined with cheap, subsidized,[6] highly digestible processed grain feeds, with rolled corn seated at the place of honor.[7] Along with growth-promoting agents, antibiotics, mechanical tenderization, and injected solutions, the feedlot diet has the effect of wiping out whatever flavor variability may have existed between cows when they walked in. The beef industry is literally gunning for "uniformity of end product."[8] That's not to say feedlot beef tastes distinctly terrible—it's what our palates are used to. What distinguishes it most is its sameness, prompting grandpas everywhere to say, "Beef just doesn't taste like it used to."

The commodity thinking that undergirds most American beef is why we found the differentiation in craft beef so thrilling and scrumptious. Within the realm of craft beef, we found tastier hamburgers (some nutty, some savory, some indecently juicy) than we'd ever had at fast-food joints—or at any nice restaurant either, for that matter. We were introduced to dozens of cuts we'd known little of previously, with names like coulotte, tri-tip, petite tender, merlot, and bavette, as well as the classically trained butchers searching them out and the chefs preparing them to gastronomic perfection. We came across farmers producing "small-batch" beef, keeping their herds intentionally small for maximum quality and flavor control, and a healthy balance between land, water resources, and livestock. And most importantly, we found something that eating grocery-store steak all your life would never tell you: No two steaks taste the same. Each offers a flavor that's a little different; each is a reflection of a million possible "ingredients" in the craft-beef equation.

Some of those ingredients:

The breed of cattle. (There are as many as 800.) What the cow ate. (And it's worth mentioning that "grass versus grain" dramatically oversimplifies the question of diet, when you consider there are small-scale farmers foregrounding millet or spent hops from beer production, or peak-season island forages or baled fermented hay. That's a far cry from the flaked corn/grass pellet dichotomy of bovine diets at industrial feed-lots.) The handling and stress levels of the cattle mattered for flavor too, as did the soil, including minerals and pH levels.

It was an equation more complex than we realized, more complex than grocery-store steak had ever had us believe.

We came to define what we were seeing this way:

CRAFT BEEF

(n) Beef produced by small-scale, independent farms with an emphasis on unique flavors and high ethical standards.

And we came to understand craft beef as the perfect storm of market demand for better beef, the now-mature local-food movement, and the wave of interest in flavor-focused, unique, small-batch foods like beer, coffee, chocolate—and even wine.

Craft beef stands on the shoulders of farmers who've already been driving toward greater "eating quality" in their herd genetics for decades, despite big beef caring for little but rapid growth rates. It builds on the work of slow-food advocates, journalists, and welfare activists who helped spread the word that industrially produced beef is no win for eaters *or* cows, and created room for an alternative market to grow. Craft beef exists thanks to the whole-animal butchers who persevered while most butchery work moved behind the supermarket deli counter, thanks to the chefs willing to serve their diners off-cuts, and even the coffee cuppers bringing a tradition of sensory analysis from one food category to another.

The main characters in this story of craft beef are farmers, butchers, and chefs, and they are united by a very simple, shared value. They treat beef—whether they're raising it, cutting it, or cooking it—as the opposite of a commodity venture. For them, it's a craft.

*   *

This book isn't a takedown of America's enormous, and enormously controversial, beef industry. That's been done thoroughly, necessarily, and brilliantly by Michael Pollan in *The Omnivore's Dilemma,* Jonathan Safran Foer in *Eating Animals,* and Robert Kenner in *Food, Inc.,* among other works. Neither is this a business tale of our internet food start-up.

What you'll find in the next pages is a series of delicious (and hopefully instructive) glimpses into the world of craft beef. Told sometimes in Caroline's voice, other times in Ethan's and Joe's, the stories and lessons that follow will teach you how humane treatment of animals is inextricably tied to uncommonly delicious steak, about the unusual and delectable cuts of beef your taste buds are missing if you never venture past the sirloin, and about the craft butchers helping reintroduce a ribeye-obsessed America to the whole cow.

We wrote this book to invite anyone who loves a good steak to journey with us into the world of craft beef, to meet the people who bring the movement to life, to contemplate the ways it improves on commodity beef, and to taste (or at least drool over) its diverse, surprising, unbelievably awesome flavors. We hope when you read *Craft Beef* that you'll fall head over heels in love with beef all over again.

*　*

The perfect place to start this story, the three of us decided over steak lunch one day, was with Caroline's first window into craft beef. Of the three of us, she was the one who had struggled longest and hardest over the ethics of meat eating. She spent almost a decade as a committed vegetarian, in large part because of the endless parade of industrial-beef exposés that populated the big screen in the early aughts. Humane treatment of animals is one of the core principles of craft beef; but when over 90 percent of American cows live on over-crowded, manure-filled feedlots, sometimes you need to see the alternative with your own eyes to believe it.

Luckily, within Caroline's first week at Crowd Cow, Joe had a solution for her quite along the lines of the *Portlandia* approach. That's where this story starts.

*   *

As Caroline walked into the office one morning, Joe told her, "I bought you a ferry ticket to the San Juan Islands. You're visiting Scott Meyers. He's got the happiest cows I've ever seen. You leave tomorrow."

CHAPTER 1

HAPPY COWS TASTE BETTER

———

OR: VEGETARIANISM, DELICIOUSLY RESOLVED

———

My ferry was late. Brigit Waring, Scott Meyers's wife, had warned me this was a possibility when we spoke on the phone the night before. The ferries that run from Anacortes to the San Juan Islands—hulking white vessels with names like *Yakima* and *Tillikum* that are part of the largest fleet in North America[1]—are frequently held back or canceled outright while operators tinker with the inevitable mechanical problems of a 50-plus-year-old fleet.

My journey to Lopez Island had begun at 6:30 that morning, when I pulled onto the northbound ramp of I-5 from Seattle, my travel mug of coffee steaming reassuringly in the cup holder. I was glad to be leaving the buzz of the city, if just for a day. I had then driven an hour and a half north, past the Mt. Vernon tulip fields that burst into radiant stripes of color each spring, and onward to the ferry terminal at Anacortes, where I now apparently had a while to wait. I didn't mind.

This was going to be my first time on a cattle farm, if you don't count sticking my head between the fence rails of a childhood friend's neighbor's yard outside of Gainesville, Florida, where red Angus cows used to graze while we played on the other side of the fence.

It was mid-August, and Joe had told me I'd be arriving at Sweet Grass Farm at the height of calving season, the time when new life begins on the same farm where bovine lives must necessarily end in order to produce some of the tastiest, most unusual steak in the world. Scott Meyers had been the first in the country to raise and finish Wagyu cattle—the Japanese breed that produces Kobe beef—on grass.

Abandoning my Beetle at the front of the line of parked cars, I made my way down to the sandy beach. I settled on one of the only logs not claimed by a perching cormorant and flipped through my copy of *The Omnivore's Dilemma*. On a dog-eared page, I read a paragraph that summed up my meat-eating conundrum, which I was hoping to resolve at Sweet Grass.

Michael Pollan wrote:

> [T]he loss of everyday contact between ourselves and animals—and specifically the loss of eye contact—has left us deeply confused about the terms of our relationship to other species. That eye contact, always slightly uncanny, had brought the vivid daily reminder that animals were both crucially like and unlike us; in their eyes we glimpsed something unmistakably familiar (pain, fear, courage) but also something irretrievably other (?!). Upon this paradox people built a relationship in which they felt they could both honor and eat animals without looking away. But that accommodation has pretty much broken down; nowadays it seems we either look away or become vegetarians.[2]

Pollan's words rang familiar. When I first watched *Food, Inc.*, I realized American meat wasn't all red Angus cows grazing happily on the other side of the neighbor's fence. I learned about the grim realities for cows at industrial-scale feedlots and realized that if I wanted to continue eating beef, I would have to perform an act of moral distancing—forget what I'd learned about the life of my steak and shove my discomfort down deep. A beautifully marbled steak was as appealing to me as it was to the next red-blooded omnivore, but I couldn't separate the deliciousness from the fact of how it got to be that way: an animal confined on a crowded feedlot, standing in ankle-deep manure while being pumped full of flaked corn.

Looking out toward Lopez Island from the Anacortes Ferry Terminal in Anacortes, Washington, on the way to Sweet Grass Farm.

Because I'd never put too much thought into the living animal behind my burger, I guess I had *already* been looking away from meat my whole life. But now, with knowledge of the industrial-beef supply chain in hand, my looking away would be more problematic. Each bite, however delicious, would leave me guilt-ridden. Each bite would make me complicit.

So I did the only thing that seemed right to me: I walked away from meat entirely.

That, however, was before I knew there was better beef out there, and long before I'd heard the term "craft beef."

Joe's assurances of Scott's animal-welfare focus in mind, I knew today might well be a day of reckoning. Could I eat Sweet Grass Farm beef—apparently the epitome of well-raised beef— without feeling like I had to look away in order to enjoy it?

*   *

I turned onto the driveway to Sweet Grass Farm two hours later, my ferry having finally docked at the forested entrance to Lopez Island several miles out into the Puget Sound[3]. As I shut the car door and took in the trees lining the clearing and low clouds above, Scott emerged from an adjacent tool shop.

We exchanged hellos and walked down the wooded drive so he could point out the cabin a short distance away, where he and Brigit live and raised their two daughters. We'd be having lunch there later, he told me. Their cabin had served them well over the years, though Scott had never meant for it to be their permanent home.

"The plan was always to build a house. Still is," he added. "But I think Brigit isn't surprised we haven't. We've been comfortable in the cabin. We just don't need a lot of material things."

At the time when Scott bought the land we were standing on, in 1999, the descriptor "farm" was generous at best. The overgrazed mudscape had been on the market for years; a former Washington State University extension agent had called it "the worst land in the county."

But then Scott had taken notice.

"It was a great chance to rehabilitate a piece of land. And nobody else was going to do it," he shrugged.

Scott, whose professional career spanned custom tool making to logging to commercial fishing, had always been interested in agriculture. His family background in orcharding and plant nurseries was what he described as "exacting work."

"When you're cropping," he said, "you're spending a huge amount of time trying to keep pests *out*. What I love about running cattle is that I don't have to be working to constantly exclude everything. Lots of other species can move through this agricultural system: plants, raptors, deer. Pasture-based cattle farming can be very inclusive."

Though the property had been damaged, in part, by the overgrazing of livestock, Scott's intensive research made him suspect that a *well*-managed cattle operation could actually heal the land.

Purebred Wagyu cows and calves graze at Sweet Grass Farm on Lopez Island.

After spending a few years grazing another farmer's cattle on the property, Scott wanted to start his own herd. USDA-approved slaughter—a requirement for legal sale of meat in the U.S.—had just become possible because the Island Grown Farmers Cooperative had devised the first mobile slaughter unit in the country. Farmers no longer had to truck their cows to the mainland for processing. Plus, Scott sensed the local-food movement had hit its stride: Western Washington consumers were clamoring for the stuff. The only question that remained was what breed of cattle to raise.

He spoke with cattle breeders from Canada to Texas and finally decided purebred Wagyu was the best choice, despite the fact that even Kobe—the most famous variant of Wagyu—was still only a blip on the American culinary horizon. The genetic propensity of Wagyu beef for marbling (those streaks and dots of fat inside the muscle) is so great that the breed was very likely, Scott reasoned, to marble well on a grass-only diet. For the past several decades, American ranchers and feedlot operators have typically relied on grain to fatten cattle up; that's why, if you ask anyone in the industry, grass finishing is considered so much harder to pull off. Scott knew Wagyu were a safer bet for the grass, and for the particular climate of the San Juans, than Angus, the favorite breed of the American beef industry.

"The San Juans and Japan are in the same Pacific Rim maritime climate, and both have poor soils," he explained. "Wagyu in Japan never had access to the best soils. That was reserved for crops."

Grass raising and finishing is definitely not how Kuroge Washu Wagyu is usually done. This most marbling prone of Japan's indigenous cattle breeds is traditionally fed a carbo-loaded grain diet and confined in pens to minimize muscle use and maximize intramuscular fat (IMF) development. What results is a steak so marbled it appears petal pink rather than red when viewed from a few feet away.

The taste of traditional grain-finished Wagyu is buttery, pillowy tender, and rich as all get-out. Joe calls it the beluga caviar of steak.

WAGYU

A4/A5 Wagyu from Japan refers to Wagyu cows with 100% Kuroge Washu breed genetics, because that's the only one of the four indigenous Japanese cattle breeds capable of achieving such high marbling. A4 and A5 are the highest quality scores awarded to beef in Japan.

Fullblood Wagyu is the term for Wagyu animals raised in America that have 100 percent Wagyu genetics and ancestry—meaning no other breed has ever been introduced into the bloodline. There are only about 36,000 fullblood and purebred Wagyu in the United States.[4]

Purebred Wagyu come from a bloodline that stems from the offspring of an Angus cow crossed with a Wagyu bull (or vice versa). Yet from that node on the family tree onward, only fullblood Wagyu paramours are subsequently introduced. One cross is bred with a fullbood Wagyu; the next, a slightly more-Wagyu cross is again crossed with a fullblood Wagyu, and so on, resulting in a higher and higher percentage of Wagyu genetics with each new generation. In the industry, the practice is called "breeding up," and the benchmark breeders must reach for an animal to qualify as purebred Wagyu is 93.75 percent or more Wagyu genetics.

Wagyu-Angus Cross is sometimes also called Wagyu cross or percentage Wagyu. It refers to cattle with at least some Wagyu in their ancestry. If you see the word *Wagyu* on a restaurant menu without further specification, you can bet what's being offered is really Wagyu-Angus cross.

Joe had also told me that Scott's grass-finished Wagyu was very different in flavor from grain-finished Wagyu but equally stunning. Gastronomically speaking, it was pioneering. The grass diet and steady use of their muscles while grazing on pasture made for a much more intense beefy flavor than you find with grain-finished beef. Food writer Mark Schatzker, who tried some Sweet Grass beef and described it in his book *Steak: One Man's Search for the World's Tastiest Piece of Beef*, called it "to steak what espresso is to coffee . . . like steak with headphones on."[5]

* *

Scott and I walked down a farm road and into the bale yard. On one side were bales of reed canary grass straw for winter bedding; on the other side was baled haylage for winter feed. Slightly fermented, the haylage smelled of vanilla root beer. During the rainy winter, Scott takes his cattle off pastures and overwinters them in a protected area with a feed bunk next to the bale yard to keep them off the wet, fragile soils and protected from the bone-chilling winds.

Environmental stressors, Scott explained, are one of the day-to-day things he monitors to keep his farm stress-free for the animals. I tucked that thought away for later.

Also in the clearing were windrows of compost; and closer to us, tucked in a secluded corner of the clearing overhung with trees, was the slaughter corral Scott had built himself. I

Top: The view out the passenger-seat window as Scott transports compost. **Above:** The bale yard at Sweet Grass Farm.

steeled myself as we walked up to it for a tour. We sure were diving right into processing—the transition from livestock to meat—within an hour of my arrival.

Here we go, I thought.

"The mobile unit," Scott explained, "backs right up onto this concrete slab."

Essentially a semitruck specially outfitted with a refrigerated trailer, carcass hooks, a butcher, and a USDA inspection agent, mobile units service farms lacking easy access to "fixed slaughter"—i.e., the kind of slaughterhouses that don't sit on eighteen wheels—and process meat directly on the farm. Scott has served on the board of the Island Grown Farmers Cooperative (IGFC), which created the first-ever USDA-approved mobile slaughter unit in the country. They're rare beasts in the world of American meat processing, which is dominated by the industrial-scale slaughterhouses that kill up to 400 animals per hour.

Of the 28.8 million head of cattle slaughtered in 2015, only 91,400 were slaughtered on farms,[6] and only a tenth of *that* number are cattle processed by mobile units. Mobile slaughter is such a small phenomenon the USDA isn't yet tallying it in its yearly cattle and slaughter inventory reports. (The 80,000 or so cattle slaughtered "on farms," but not with a mobile unit, are likely attributable to custom slaughter, which operates under state, rather than federal, inspection. That's generally how cow-pooling among a couple friends works.) Scott feels fortunate to have access to a mobile unit, despite the investment of money and labor hours it required to design and maintain.

Scott opened the corral gate into the main holding area, and we stepped inside. I noticed small purple flowers growing up against the sides of the steel panels. Underfoot, the grass was springy, dark green. Branches scored the sky above us.

He pointed out each of the slaughter corral's most important features:

- The high-sided slaughter corral panels create a visual barrier, preventing the animals waiting their turn from seeing what's happening ahead.
- The deep bedding of fresh straw provides comfort and stress reduction, as well as keeping the cattle clean.
- A drainage system allows blood and wash water to be pumped away and recaptured in compost piles, ensuring nutrients are cycled and—just as importantly—the cattle in the holding pen can't smell the blood.
- "Companion steers" that won't be slaughtered are brought into an adjacent pen where they can touch noses with the steers in the slaughter pen. "They're herd animals," Scott explained. "You never want the last one to feel like it's alone."

"All of this," he said, motioning around us with an arm, "is thought out and designed to keep them calm. You don't want adrenaline before slaughter."

He was referring to an idea that was new to me: that animal stress directly impacts the taste of a steak. I had always assumed that humane treatment was a cost of doing business— something decent farmers would practice out of kindness to the animal, but disconnected from the function of raising profitable protein. Scott, however, offered a new perspective.

"Having calm animals," he said, "has everything to do with the quality of the beef."

* *

When I looked to the internet to understand the relationship between stressed-out cows and bad-tasting beef, I found the phenomenon to be well documented.[7, 8, 9, 10] Some animals that are highly stressed out can even produce meat that is so dark it looks bruised—that's when a carcass qualifies as a "dark cutter,"[11] is tough and dry, and goes bad more quickly than usual.[12] Dark cutters accounted for 1.9 percent of the cattle commercially slaughtered in America in 2016.[13] Although 1.9 percent might not seem like a lot, it's nearly 600,000 cattle.[14] (And it doesn't include the approximately 2 percent of feedyard cattle that die on the feedlot each year due in large part to disease and damaged lungs from breathing in too much manure.[15])

Slaughtering on-farm, like those that use mobile processors do, gives farmers quality control right up until the last moment of the animal's life. When cow-calf producers sell their calves into the commodity system, the remainder of those cows' lives—including conditions at slaughter—are unknowable. That's a problem because stress levels are intrinsically tied to quality of meat.

COW-CALF PRODUCERS

The vast majority of cattle ranchers in the U.S. are cow-calf producers, which means they raise calves for a living to be sold to feedlot buyers. Those feedlots in turn 'feed out' the cattle on concentrated diets as quickly as possible.

"The cow-calf producer," Scott said, "wants genetics that produce strong, fast-growing calves. It's not about the culinary outcomes, or the taste. The feedlot never gives the cow-calf producer information about the eating quality of the beef, so there's nothing for him to base his genetic selection on when he breeds next year's calves. It's impossible to improve the eating quality of U.S. cattle genetics when there's that disconnect."

Apple trees dot the mama cows' paddock.

Genetics play a role in stress—Scott will remove an animal permanently from the herd if it has an innate tendency to spook or be spooked and can't be trained out of it. The last moments before slaughter, too, are a vital part of the stress equation. Yet perhaps most important in the puzzle of nature and nurture as it relates to animal stress and meat quality is the *overall* quality of a steer's or a heifer's life. A key component of Scott's farming program—and how he ensures he is raising a premium product—is creating an environment for the cattle that is naturally stress-free.

One component of the stress-free environment is Scott's own behavior, from attuning himself to "the bovine language of body interactions" to building farm infrastructure that works as smoothly as possible, ensuring he never has to "fumble around" while he's working near the herd, which might startle them.

"All of that," he explained, "is about getting the best meat that these genetics have to offer."

*　*

I was pouring over these thoughts as we opened the door to Scott and Brigit's cabin, where Brigit was just taking a cast-iron pan off the stovetop. In it were three purebred Wagyu steaks we would be sampling: the tenderloin, the ribeye, and the striploin—considered the three most premium cuts on the entire cow. A cold greens salad, along with a hot dish of beets and potatoes dyed magenta by the steaming roots, waited for us on the round oak table. (Incidentally, Joe told me later, these dishes resembled the typical pairings for Wagyu in Japan: leafy greens and root vegetables.)

As Brigit told me about the origin of the vegetables we were about to eat ("we get them fresh from our neighbors, Ken and Kathryn"), I looked around the small first floor.

The two-story cabin was full of light; some walls were paneled in tongue-and-groove pine. Against one wall sat a small, ivory-colored Monarch cook stove with sea-green accents. The space was not roomy but extremely homey.

Brigit forked the three steaks onto a wooden cutting board, sliced each into strips, and told Scott and me to dig in.

I sliced off a strip of the rib cap—the most marbled part of the animal—and bit down.

It was juicy, as soft in texture as ricotta gnocchi, and much richer than I expected for a grass-finished steak, which are generally known for being leaner than their grain-finished counterparts. The richness came from those legendary Wagyu genetics. Yet this was not the mild, buttery creaminess of traditional grain-finished A5 Wagyu from Japan. There was some type of nutty, earthy, and tangy tone—probably from the grasses—underlying the flavor; it was deeply beefy.

It was also pretty extraordinary in terms of nutrient profile. Grass-finished beef generally gets crowned healthiest steak because of its low ratio of omega-6 to omega-3 fatty acids. Western diets are far too high in omega-6 fatty acids (because of all the processed grain we consume), with a ratio of n-6/n-3 in our diets around 15:1.[16] Evidence suggests the ideal ratio is closer to 1,[17] and that's exactly where grass-finished beef ranks because of all the linoleic acid transferred from grass into the meat of grazing animals.[18] But the issue is complicated, because it turns out that what the cow eats isn't the only influencing factor on the nutrient quality of its meat. Breed matters too. Wagyu beef, compared to Angus, has much higher levels

Two grass-finished purebred Wagyu tenderloins from Sweet Grass Farm.

of unsaturated fats (the good ones). How easily the fat on a steak melts is a quick-and-dirty proxy for determining how good it is for you. Extremely firm fat on a steak tends to indicate you're looking at saturated fat—not exactly known as a silver bullet for health. In the case of Wagyu beef, the presence of so much oleic acid—a monounsaturated fat—causes the fat to melt against your hand if you so much as touch it.

As I chewed, Brigit chuckling at my blissful silence, I heard a cow mooing from its paddock a few hundred feet away. I was surprised to find that the thought of the live animal nearby, *as I chewed one of its cousins*, didn't particularly bother me. The reason had everything to do with the setting, and with Brigit's and Scott's attitudes. What I'd seen of their behavior so far had been reassuring—they were purposeful in their role as

CALLING A COW A COW

To a rancher, not all cows are called cows. We hope you'll forgive that in this book, we use the terms rather loosely. But if you are to become a true beef nerd, the following distinctions are worth knowing:

A cow is a mama cow. Farmers who raise cattle from birth all the way to slaughter typically divide their herds into a few subgroups, one of which is the cow-calf herd—the mothers and babies. They are kept together until the calves are weaned and graduate into the "feeder" group. If farmers name their cows, it's typically only the mamas that get christened.

A heifer is a virgin female cow who will either be raised for meat or eventually be bred with a bull, get pregnant, and in so doing, earn the moniker "cow."

A steer is a castrated male cow who will be raised for meat. Farmers, Scott included, typically don't name their steers, because the added emotional attachment makes slaughter day too difficult.

A bull is an uncastrated male who will typically sire most of the new calves in a herd each season. Farmers choose their bulls carefully—sometimes owning them, sometimes renting them from another farmer—for a variety of traits they'd like to see passed on in the next season of calves.

animal caretakers and determined to strike a symbiosis on the farm between soil, livestock, environment, and human behavior. They were exactly who you'd want to produce your meat.

But I still hadn't seen the cows for myself.

As I was taking another bite, the landline phone rang. Scott picked up and spoke into the receiver. I heard a voice crackle on the other end of the line but couldn't make out the words.

Scott replaced the receiver on the wall, grabbed his flannel overshirt off the back of his chair, and said to Brigit: "That was Kathryn. Apparently we've got a cow ready to give birth."

He turned to me: "Want to see it?"

He strode toward the door. "Bring the steak."

So I followed him out the cabin door, slice of premium Wagyu beef in hand, to go watch the birth of a future steak. The irony was not lost on me.

We headed through the backyard and toward a paddock that held 100 animals: pregnant mothers and new calves. (The feeder steers, in the process of fattening up on forage, were a ways off, on the other side of the farm.)

We found the cow in question and crept toward her, stopping 15 yards shy. Scott took a pair of binoculars I hadn't noticed him don from around his neck and handed them to me so I could get a closer look. We were staying a respectful distance to ensure we didn't cause the mom undue concern—another small but important piece of the puzzle in keeping every stage of the herd's life naturally free of stress.

When the calf finally came out, it came quickly. We watched, waiting to see how it would adjust. I imagined it would be a shock to enter from a very dark, warm space into the bright, dry afternoon sunshine in which we stood. Around us, bugs buzzed. The mother stood, turned, and started lowing in a rough, sweet tone.

Scott said, "That's a sound only mother cows make to their calves. That's a mother sound."

He and Brigit shared a quick smile. Though this was a regular occurrence—the second calf in 24 hours, I was told—it was still noteworthy.

"These are special moments," Scott said.

"Now just watch," Brigit said.

Scott added, leaning toward me, "The mother cow will lick that calf to standing in the next five minutes. That tongue may look soft from here, but it's rougher than sandpaper. It'll get the calf's blood flowing."

A new calf on Sweet Grass Farm.

And indeed, the cow continued to low, between busy licks; and the calf tried to stand but fell. Again it stood on the two front legs, the last ones not quite able to make it up. It fell again and again, and then finally stood.

Brigit put her arm around Scott.

I remembered suddenly that I was still holding a strip of steak in my hand. Watching this birth, I thought I should feel a little strange eating one of the calf's older cousins, but it felt oddly okay. As Scott and Brigit smiled upon their herd, I could see that the animals lived good lives and died quickly and respectfully. Brigit had a freezer full of beef yet was able to keep another foot squarely in the world of caring for and loving the animals she helped to raise. This was the omnivore's dilemma. At best, I was a student attending a lecture *about* the omnivore's dilemma. Brigit and Scott, these producers of small-batch, craft beef, straddled the dilemma effortlessly every day, like farmers and all people who raise their own food always have.

As I bit down on the Wagyu steak I was holding in my bare hand, I noted the richness of the fat, as well as the tang present in both the flesh and the fat itself. It tasted a bit like the air out there smelled, the hot sun baking the grass and lifting some nutty scent off the seeded tops. It had the texture of a cloud. These animals probably had some of the least stressful lives of any living creature on the planet. If all beef was as well raised as this, I could be an omnivore.

Sunset over the San Juans on the ferry ride back to Anacortes.

CHAPTER 2

SELECT—CHOICE —PRIME—CRAFT

——

STEAKHOUSES ON THE CUTTING EDGE

——

For decades, the go-to standard for steak in America has been corn-fed Angus, USDA Prime beef (from a feedlot). But the explosion of interest in organic and grass-fed beef in the past few years hints that conventional wisdom about what makes for a great steak is fraying at the edges—that the American eater is looking for something different. To find out where tastes are headed, I ventured into the belly of the beef: America's finest steakhouses.

The leather-bound lunch menu at Delmonico's in Manhattan, New York.

DELMONICO'S: THE NEW THRILL OF OLD MEAT

My steakhouse tour, Joe told me, had to begin at Delmonico's. As America's oldest restaurant, the Manhattan eatery holds a hallowed place in the hearts of steak lovers everywhere. Proudly retaining gold-filigree vestiges of its 1800s grandeur, the original home of Eggs Benedict and Lobster Newburg struck me as a place likely to take a conservative approach to steak.

But a little research told me otherwise. Chef Billy Oliva has spent the past five years taking the Delmonico steak in an entirely new, old direction: extreme dry-aging.[1]

When Chef Billy joined me for an after-lunch coffee at a table by one of the dining room's expansive windows, I was just emerging from a mind-addling steak dreamscape.

The 45-day dry-aged rib steak at Delmonico's.

I had just polished off the last bite of a 45-day dry-aged bone-in ribeye roughly the size of Massachusetts. Only the long rib bone, which stuck out from the meat like a saber, remained. I felt a mix of regret and disbelief that I'd vanquished so formidable a steak. I would have greeted a second round with enthusiasm.

Its aged flavors (quite unlike the relatively bland flavor of the unaged steak I'd eaten at another prominent New York

THE ART OF DRY-AGING

The modern history of dry-aging starts with Marie-Antoine Carême, the father of France's *grande cuisine*, who is credited with bringing French food into global focus.[2]

Carême knew, as humankind surely did for millennia prior, that dry-aging a side of meat in a cold, low-humidity room will preserve it. (Do it wrong, and you'll end up with rotten, inedible meat. Dry-aging proponents speak proudly of their craft as "controlled rot," *control* being the operative word.)

But Carême wasn't preserving meat for preservation's sake alone.

By the 1800s, French chefs had begun noting that—not unlike what happens to cheese when you let it rest in a cave for a while, developing colorful, fuzzy molds—time seemed to do something strange and magical to the taste of beef.

On a scientific level, steak loses moisture as evaporation occurs, and it becomes more tender as fats are oxidized and natural enzymes in the muscle cells soften connective tissue[3] and break proteins, carbs, and fats down into amino acids, sugars, and fatty acids.[4] The amino acids, for their part, produce an umami flavor, and the breakdown of carbohydrates creates a sweet taste.[5, 6, 7]

steakhouse the night prior) had struck me dumb. In the 45-day dry-aged steak there was a certain fullness you don't get in fresh beef, a roastiness that lingered for minutes after I swallowed, a sharp bite of caramel and carbon from the seared crust.

Carême, known as a bit of a crackpot among his staff and patrons (he was innovating, after all!) insisted dry-aging be allowed to go on "as far as possible."

Dry-aging has experienced a resurgence of popularity in the past few decades, particularly at high-end restaurants and butcher shops.[8] But relatively speaking, it's still rarely seen. What's far more ubiquitous is "wet-aged" beef. Ninety percent of the beef at American grocery stores, in fact, is wet-aged,[9] which is a new thing that developed in the 1960s. More on that in a moment.

The question I was dying to ask Chef Billy—the question, in fact, that I'd been pondering since I first heard about the revived art of dry-aging—was this: Can you actually taste the difference between a steak dry-aged for a week and one dry-aged a few months, like the one I was eating?

"Absolutely. Dry-aging is a craft where you really can control the process."

Those who know what they're doing, he went on to explain, are able to coax increasingly unusual flavors out of the beef over time: "blue cheese, roasted hazelnut, and mushroom."

Chef Billy's foray into dry-aging began in 2012 when he added his first dry-aged steak to the menu. Today, the menu is positively marbled (sorry, had to) with offerings of dry-aged beef, including a commemorative 180-day dry-aged steak honoring

AGING: THREE TERMS YOU HEAR

DRY-AGED

Dry-aging is a traditional method of hanging primals or whole sides of beef in a refrigerated room to dry for a period of a few days up to several months, during which time enzymes get to work tenderizing and altering the flavor of the beef.[10] Dry-aged beef has many ardent fans who champion its effect of creating more tenderness and "concentrated flavor." It comes with a heftier price tag than its unaged counterpart, because waiting weeks to move saleable product is costly for butchers and restaurants.[11] Dry-aged beef also loses water weight as it evaporates—6.53 percent after 21 days, according to one study.[12] That might not sound like much, but even going from an initial water weight of about 75 percent down to 70 percent means the juices that remain are more concentrated, more flavorful.[13] In addition to water loss, the crusty (sometimes moldy) exterior of the dry-aged primal must be trimmed (up to another 6.55 percent after three weeks).[14] Both of those are losses the butcher must ultimately recuperate in the cost of the steak.

Many connoisseurs explain dry-aged steak this way: It's like a Bordeaux of premium vintage compared to its just-juiced cousin. Dry-aged beef tastes nutty, roasty, sweet, and meaty.[15, 16] And until about 50 years ago, this was the norm.[17]

WET-AGED

The development of vacuum packaging in the 1970s made it possible to transport beef faster and farther across the country, and to give it a longer shelf life in the grocery store. Cryovac wrapping sealed juices in and kept air out during the week-long trip from packer to supermarket, and some savvy marketer decided to give this economically convenient period of steak dormancy a traditional-sounding name: "wet-aging." In reality, it's more supply-chain improvement than ancient culinary technique.

The science tells us that wet-aging, like dry-aging, does in fact make steak more tender, but its effects on taste are less advantageous. One study found that wet-aged beef produced bloody/serumy and metallic flavors (compared to dry-aged beef's more pleasant description: beefy, brown-roasted.)[18] Other studies conducted by trained sensory panelists agree that wet-aged beef is strongly serumy and sour; though still other studies were less conclusive. The authors of a big review on this controversial topic settled the matter this way: If you're familiar with wet-aged steak—as most Americans are, since 90 percent of grocery-store beef is wet-aged[19]—it won't necessarily bother you.[20] Extrapolate a bit: Once you've had dry-aged, you don't want to go back to wet-aged.

FRESH

Fresh beef isn't aged at all—it's sold within a day or two of being cut, and as a result, it can sometimes be tough. A really well-raised cow free of stress and fed a high-quality diet, however, should be able to produce at least some fairly tender cuts of beef without any dry-aging.

The biggest beef-eaters in the world[21]—the people of Argentina and Uruguay—prefer their steaks fresh, citing a stronger beefy flavor and not minding the added chewiness. The method of slow roasting over the indirect heat of *carbon de quebracho blanco* (a charcoal) probably helps tenderize and alleviate toughness somewhat, but it's also likely a matter of taste. There are different texture preferences from one food culture to another.[22]

the restaurant's 180-year anniversary. In 2018, Chef Billy will be introducing a stand-alone dry-aged menu with steaks ranging in age from 90 to 180 days.

"We're really going for full-blown art," he explained.

And it's starting to turn heads.

"Just last week," he went on, "A French couple came in who had made the trip to the restaurant just to taste the 180-day dry-aged steak. Usually in France it's only 21 to 28 days max of dry-age, so more and more people like them are hearing about what we're doing and seeking us out."

He might even move beyond steak:

A dry-aging course Chef Billy took in Asia exposed him to other dry-aged meats and even vegetables, like cauliflower. I told him I'd return soon to check on his progress and try a 180-er. By then, my beefy, roasty, 45-day dry-aged ribeye might look like child's play.

AT 212 STEAKHOUSE, ALL EYES ARE ON KOBE

In the past few years, A5 Wagyu imported from Japan has swept the highest end of America's steak scene thanks to its unique umami flavor and unearthly level of intramuscular

DRY-AGING AT HOME

Increasingly, it's the mark of a true steak aficionado if you dry-age your own beef. Dry-aging at home basically involves setting aside a roast or a whole primal (one of the chunks like the sirloin or the rib into which the butcher subdivides the cow) in a dedicated refrigerator for up to several weeks (with temperature and humidity held as constant as possible to prevent fruity-green molds from sprouting) to increase tenderness, flavor, and the funky aromas that dry-age connoisseurs so adore.

Two of our most trusted beef authorities—Steven Raichlen and J. Kenji López-Alt—recommend many of the same elements for basic setup,[23, 24] so we combined and summarized their tips for at-home dry-aging:

- First, you really should arm yourself with a **separate refrigerator**. Got a mini-fridge lying around in your garage? That works great. The danger of plopping a steak right in your normal fridge is odor exchange. The raw meat will absorb all the Brussels sprouts, chocolate cake, hard-boiled egg, and tomato soup smells, and vice versa. Not good.

- Next, **get a whole, honking primal, or at least a full roast**. You can't do individual steaks, because of surface area (you'd have to trim off too much of the steak). Start big.

- Then you'll need **a wire rack and a tray**. That's for max airflow and minimum meat-on-tray contact. If you plop it directly on a plate, you'll get a piece of meat that's fully rotten on the bottom after as little as a week.

- Finally, **wait**. As little as a week creates added tenderness; after that, patience is increasingly rewarded with funk.

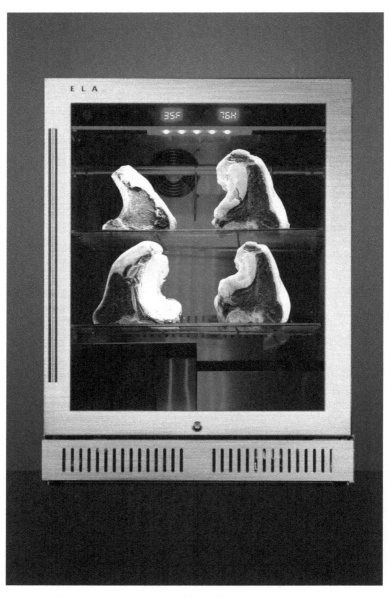

Primals of beef aging in at-home dry-aging fridge Steaklocker.

marbling—a trait distinctive to the Wagyu breed Kuroge Washu, which gives rise not only to the Kobe brand but to other high-reputation Japanese brands like Matsuzaki and Ohmi. Though Kobe is much more famous in the United States, Ohmi has the longest history as a premium brand inside Japan.

So why is Kobe more famous in the U.S. than any other type of A5 Wagyu? The reason boils down to an ingenious marketing campaign by the Kobe Beef Marketing & Distribution Promotion Association.

You'd be forgiven if you think of Kobe as old hat—it's basically ubiquitous at steakhouses in most midsize cities in America, right? Wrong. A mere 25 restaurants in the entire U.S. are certified to sell Kobe beef, so the "Kobe" you encounter at almost all American restaurants is fake.[25] Real Kobe beef must meet the following conditions. It must:

1. come from the Kuroge Washu breed;[26]
2. be from a cow that was born in Japan's Hyogo Prefecture (sometimes you'll hear people say, "It has to come from the Tajima genetic strain, which basically means Kuroge Washu that was raised in Hyogo); and
3. achieve a carcass quality score of at least A4.[27]

Like Champagne, Kobe beef is protected by a marketing tool called geographical indication. Just as you can't call a sparkling

和牛

What is Wagyu?

Wagyu (和牛 - pronounced /WAH-gyoo/) is a term that literally means "Japanese cow" and is the name given to cattle breeds developed over centuries in Japan. Wagyu beef is known for its intense marbling and carries a well-deserved reputation for exquisite taste, texture and tenderness.

Genetics and Breeds

The world of cows in Japan can be traced by breed, bloodline, and even geography. Beginning in the Meiji Era, the Japanese recognized the value of Wagyu beef and sought to develop it into the highest quality beef in the world. Seeking to preserve the purity of the bloodline, Japan banned the export of Wagyu DNA in 1997.

There are four main breeds of Wagyu in Japan:

和牛

Japanese Black 黒毛和種 **Kuroge Washu**
80% of cattle in Japan

Japanese Brown 褐毛和種 **Akage Washu**
18% of cattle in Japan

Japanese Shorthorn 日本短角種 **Nihon Tankakushu**
1 - 2% of cattle in Japan

Japanese Polled 無角和種 **Mukaku Washu**
< 1% of cattle in Japan

A5/A4

By far the most important of the four Wagyu breeds is **Kuroge Washu**, known for its unique genetic disposition for exquisite marbling. Kuroge Washu is the only breed of cattle that can achieve an A4 or A5 rating by the Japanese Meat Grading Association.

Kobe Confusion

You've probably heard of Kobe beef. But thanks to mislabeling in the U.S., the term is subject to much confusion. For example, beef that is actually only a small percentage Wagyu is often sold at a high premium and labeled as "American Kobe."

In order to be certified Kobe beef, the cattle must be Japanese Black (Kurogewashu) and pure Tajima. Also, the cows must be bred, raised and slaughtered in the Hyogo prefecture of Japan. Finally, the meat must reach a 6 or above on the BMS marbling scale.

Marbling and the BMS Grading System

Marbling is the distribution of soft white intramuscular fats within the red meat. Generations of careful breeding and management of diet and exercise contribute to the unique marbling of Wagyu Beef. Wagyu is graded for its marbling on a scale of 1-12, with 12 being the highest achievable quality. At this level, the beef is considered a work of art!

To be designated as "A5 Wagyu," the beef must be raised in Japan and achieve the highest possible rating by the Japanese Meat Grading Association. For reference, USDA Prime beef, the highest designation of quality in the U.S., is equivalent to a 4 or 5 marbling score.

USDA Prime
Highest quality beef available in the U.S.

BMS 1 | BMS 2 | BMS 3 | BMS 4 | BMS 5 | BMS 6
BMS 7 | BMS 8 | BMS 9 | BMS 10 | BMS 11 | BMS 12

A5 Wagyu
Beef with an "A" yield rating and a BMS score of 8+ earns the title of A5, the highest designation of beef.

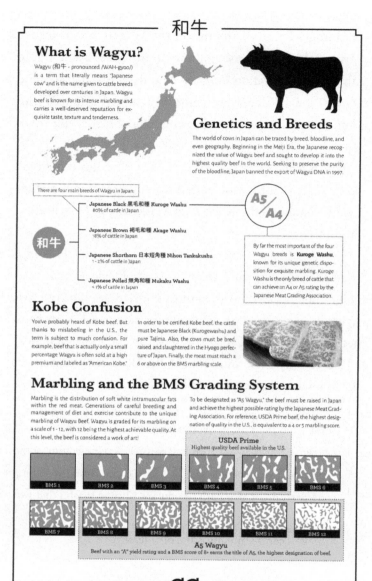

CC
CROWD COW

white wine grown in California "Champagne," you can't call Kuroge Washu cattle raised on a Nebraska feedlot "Kobe."

But enforcement of the rule is uneven. While Japan and the European Union have a trade agreement preventing knock-off Kobe from sneaking its way onto opportunistic restaurant menus, the same stringency isn't applied in the United States. That's why you can order "Kobe sliders" at your average midlevel pub in Nashville, Pittsburgh, or Seattle at a price two to three times that of an average burger but nowhere near what real Kobe would cost (that's part of how you can sniff out the fakes—a half-pound burger with real Kobe ground beef would cost well north of $100). If you're looking for authentic Kobe, you're best off sticking to those 25 certified restaurants (see Appendix I at the back of the book)[28]—or ordering online.

In reality, the beef labeled "Kobe," "Wagyu," or "American Kobe" at most restaurants across America is a cross breed of Wagyu and Angus cattle.[29, 30] (In Australia, bless them, they call it "Wangus.") Wagyu-Angus should be appreciated on its own terms: for the greater-than-Prime levels of marbling it can achieve and for the deep beefy flavor it develops while roaming on pasture—a flavor notably absent in A5 Wagyu from Japan. While A5 Wagyu or Kobe is a completely different beast from traditional American steak, Wagyu-Angus can be thought of as more like a familiar American steak, elevated.

And that's exactly its point. Starting in the 1980s, a group of American ranchers spearheaded an all-out Wagyu-Angus campaign to develop the cross, which they thought would be the perfect steak for the American palate: marbled beyond even USDA Prime levels but not so intensely decadent and rich that you have to limit yourself to indulging in only a few ounces at a time. If your tune is to polish off 16 or 32 ounces of steak in one sitting, you can do it with Wagyu-cross. You might destroy your body if you tried that with Kobe.

* *

While in New York, I decided I should hear from the owner of one of the few restaurants in America offering authentic Kobe from Japan: 212 Steakhouse.

When I walked into the entryway on a warm summer evening in Manhattan, I passed an ostentatious golden statue of a cow's head. Nikolay Volper was at the bar, chatting animatedly with a group of suit-clad guests, but he tore himself away to say hello.

A Bulgarian-American of towering stature, Volper had a heavy accent and an infectious enthusiasm for all things Kobe.

He halted our pleasantries to lead me back toward the golden cow, which, he told me, was the proof of authenticity handed

A5 Kobe sits in front of a plaque certifying its authenticity.

out by the Kobe Beef Promotion Association.

Like Chef Billy at Delmonico's, Volper had picked a focus and was swinging all his energy toward it.

He offered Kobe for a simple reason, he said:

"To be a steakhouse these days, you need premium beef. And premium means real, Japanese Kobe."

For Volper, it's black and white. The melt-in-your-mouth feel—thanks to the low melting point of Wagyu meat[31]—combined with the filigreed sprinkling of fat across the muscle render this most marbled of Japanese beef the ultimate offering for a high-end steakhouse. And he's not the only one who feels that way.

A5-ranked Kobe sears on a pan in Japan.

While 212 Steakhouse offers the Kobe ribeye, striploin, and tenderloin at $25 an ounce each (with a three-ounce minimum), other locations demand heftier sums for the prized meat. When the Wynn Las Vegas offered it recently as a special, four ounces of authentic Kobe set diners back $220, with each additional ounce going for another $55.[32]

Given his insistence that Kobe is the gold standard, I wondered what Volper thought about the viability of the steakhouses that focus solely on USDA Prime.

He told me he doesn't worry for them. On his own menu, he still offers USDA Prime filet mignon (a small "filet" of the tenderloin muscle) and New York strip, along with Australian Wagyu, which is Wangus raised in Australia. But he does think the refusal to branch out past Prime isn't doing those other

COOKING KOBE OR A5 WAGYU
TO PERFECTION AT HOME

If you've managed to get your hands on some authentic Kobe or A5 Wagyu steaks, you might be sitting at your kitchen table with some frozen steaks in front of you, wondering how to go about not ruining them. If they were the real thing, after all, you just paid an arm and a leg for them. Here's what Joe recommends you do.

- **First, invite friends.** Kobe and A5 Wagyu steaks are so rich that they're best enjoyed in small portions, among friends. By no means do you need 32 ounces each.

- **Defrost right.** Place the cuts you're planning to serve in your fridge to defrost—this can take up to 48 hours. *Do not* defrost by placing them on the counter. The melting point of fat in Kuroge Washu beef is 77 degrees, which means the marbling (a.k.a. fine ribbons of intramuscular fat) in your steaks will literally melt away before you get them into the pan. Floridians, take heed. Fridge-defrosting is the way to go.

- **Slice into bite-sized portions.** After removing steaks from the fridge, slice into small strips that you'll cook one at a time. Remember: Small portions!

- **Season conservatively.** The purpose of seasoning is to help a steak sing. But Japan's most marbled Wagyu beef makes its own music, so season just with a bit of high-quality salt to help bring out the sweetness and umami.

- **Cook on stainless steel.** Sear each slice directly on the hot surface of a pan for one to two minutes per side. No oil or butter needed—the steak itself holds plenty and will cook in its own fat.

steakhouses—he called out one famed Brooklyn institution by name—any favors.

Without Kobe, he said, that restaurant simply "cannot be the best steakhouse."

A5 Wagyu from Japan's Kagoshima Prefecture sears with vegetables on cast iron, in the traditional small strips typically scene in Japanese cuisine.

Larry Olmsted's book *Real Food/Fake Food* has helped Kobe build an even larger following in the past two years, sending a huge wave of customers to 212 Steakhouse.

"Most people who walk in these days," Volper explained, "are here for the Kobe from Japan. Interest in it is just going to grow more and more."

THE LONG REIGN OF USDA PRIME

If you're a steak buff, what you've been told your whole life is that Prime is as good as beef gets—a framing that ignores other quality metrics like the breed of cattle (there are as many as 800 in the world!), the quality of the feed, the living conditions of the animal, and on and on. How good or not good a steak tastes is much more complicated than the USDA would have you think.

Prime sits at the top tier of the USDA beef grading system, followed in descending order of quality by Choice, then Select, and below that, Standard or Commercial, and below *that*, Utility or Cutter. The latter two grades are so tough and low quality that they aren't usually even sold at retail except as supermarket ground beef.[33]

But what exactly does Prime mean? Here's where you have to understand marbling. The entire system of meat grading in the United States is predicated on it. When USDA meat graders look at a carcass of beef, they're scanning for two factors: physiological maturity (they like 'em young, less than 30 months of age) and marbling.[34, 35]

If you're standing in the grocery-store meat aisle, picking out a steak, you're probably looking at a crimson piece of meat, perhaps rimmed by a layer of white fat. If you bring the steak a little closer, you might notice tiny grains of white or cream—depending on the steak's grade—interspersed throughout the red meat. For Select-grade meat, you might see an expanse of uninterrupted red. For Choice, you'll notice a fair few streaks or flecks of white. At Prime, you'll see a reasonable-to-high amount of the creamy marks. That stuff's marbling. Its technical name is intramuscular fat, and it has long been the pride and joy of American steakhouses, pitmasters, and people with expense accounts everywhere. Businesses have been born and have crumbled over it (probably).

WHY MARBLING ISN'T EVERYTHING

But in recent years it's become clear that marbling isn't everything. Food writer Mark Schatzker has reported on the nascent literature, which hints that steak flavor—rather than depending on marbled fat—in large part derives from phospholipids (fats invisible to the naked eye), along with the range of organic compounds that seep into a steer's muscles as it grows older.[36] Those organic compounds, which still aren't well understood, are what create that difficult-to-describe "beefy flavor" that distinguishes beef from the meat of other grazing ruminants, like buffalo or venison.[37]

Marbling is good at creating both mouthfeel and a taste we could describe as richness. Think of what happens to a bowl of plain pasta noodles when you add a pat of butter. Now those noodles look a whole lot more edible, don't they? We're hardwired to interpret fat as reward.

But (!) the butter isn't the whole story of how that pasta tastes, is it? The noodles could be overcooked, or just plain bland, and even the richness of butter can't fully salvage that. Now, what if you had handmade pasta dough with homegrown semolina wheat and the perfect balance of egg yolk and salt, cooked *al dente* in boiling water? You'd be tempted to eat that pasta naked, no butter or marinara, straight from the colander.

The analogy is oversimplified—as with butter, the nutrient profile and flavor of marbled fat can vary depending on what the cow ate. Not all intramuscular fat is created equal, so it can affect the overall flavor profile of the steak to different degrees. But the point remains that as with wine, myriad factors can affect the taste and tenderness of a steak: breed, feed, age, and handling by both the farmer and the butcher. To rate its overall quality by only one characteristic—the volume of fat speckled inside the muscle—is to miss the forest for the trees.

The chalkboard at Bateau that lists the day's cuts.

BATEAU: TURNING THE STEAKHOUSE ON ITS HEAD

Steakhouses like Bateau in Seattle are starting to buck the long-lived Prime tradition, taking their menus in an entirely different direction.

One summer afternoon in Seattle, I sat down with Taylor Thornhill, the chef de cuisine at Renee Erickson's restaurant Bateau, which food critic Bill Addison called Erickson's "magnum opus" when he recently named it one of America's 38 best restaurants.[38]

Unlike the vast majority of steakhouses, which receive a continual, wholesale supply of just the few most popular cuts of beef, Thornhill and Erickson receive one grass-fed, grass-finished cow per week from Erickson's Whidbey Island farm, La Ferme des Ânes, along with beef from a few other farms with similar values. They also butcher and dry-age in house and tweak the menu offerings as various cuts sell out.

The concept is revolutionary, and the design mirrors the code-switch: Rather than the clubby, smoke-filled interior of a typical steakhouse, the airy bistro I entered was filled with sunshine and painted white.

On one wall, a chalkboard displayed the dozens of cuts of beef available that Saturday and sported an intricate chalk sketch of a cow's head. In a far corner inside a windowed room, a bundled-up butcher worked to break down a side of beef, his breath coming out in puffs.

Thornhill explained that the breeds of cattle raised on La Ferme des Ânes and served at Bateau—Charolais, Maine-Anjou, and Limousin—offer the diner "something different" than American Angus.

Bateau's decision to showcase unusual continental breeds of cattle is pioneering, and Thornhill explained why.

The commodity beef industry, he said, "identified one specific breed of cow"—by which he meant Angus—"that does really well in terms of growing really fast and being low maintenance,

CONTINENTAL BREEDS

Most of the steak in America is Angus (or Angus-cross), which is a British breed. Continental breeds refer to those that come from the rest of Europe—Continental Europe. Below are some of the most popular and common of the Continental breeds.

Charolais

Known for a general coarseness[39] balanced by good marbling and buttery taste,[40] the cream-colored Charolais breed hails from Southeastern France in the region of Charolles and Nièvre, where it is still much loved for its flavor.[41] According to legend, by the ninth century, Charolais were prized by the French for their draft power and sheer muscle.

Maine-Anjou

For cattle being raised on a grass diet, Maine-Anjou produces among the best-marbled meat.[42] This breed, typically brown and spotted with white, comes from the Northwest region of France. Famous for their muscularity and hulk-like size, Maine-Anjou cows grow to 1,500 to 1,900 pounds, and the bulls can get as large as 3,100 pounds. In 1843, the agriculturalist Leclare-Thouin wrote that in the Auge Valley grasslands, the Maine-Anjou's ancestors fattened fastest. They were "the last to be put onto the grass, but were the first to be picked out to go to the markets in the capital city."[43]

and that all goes into profits for the people growing them. On the industrial side, they don't let cows live longer than 12 to 14 months."

At Bateau, by contrast, they slaughter their cows at "well over 30 months," because of the greater flavor that develops with age, and embrace the challenges that come with using non-Angus breeds.

"Sometimes we'll have a beef that's super lean," Thornhill explained, "and we have to help consumers understand that what they're eating will be a little different."

Not everyone is adventurous enough to try the five-course tasting menu, which depending on what's available, might showcase beef belly, heart, liver pâté, or a beef-fat-based dessert, but the sheer variety of cuts they have to work with gives them—and diners—some wiggle room.

Provided the kitchen hasn't run out, you can order a ribeye or a New York strip steak, or expand your palate with a dish of sweetbreads or carpaccio cut from a challenging part of the cow, the mock tender (a.k.a. chuck tender).

"Despite its nomenclature," Taylor said, "the chuck tender is not tender at all. At first we tried to cook it as a steak. That," he laughed, "did *not* work."

"Next we tried bresaola-style curing, which is air drying, and that was okay . . . but it didn't have a ton of flavor. So we ended up curing it very far and shaving it very thin and doing a broth."

But even that method didn't last.

"Finally," Taylor said, "we just started making carpaccio out of it, and that's worked out pretty well. We cut it into thin strips and dust it with burnt onion ash so it turns completely black, and we use this thing called Activa, which is a meat glue.

"We bond it back together, and it creates this marbled effect— you have veins of black all the way through, so when you slice it, it's red marbled with black."

The many-step process is worth it in the end.

"Everyone who gets it thinks it's great. And it's really exciting for us, because it's a completely new way to look at carpaccio."

* *

For me, the tasting menu was a must, so I came back in December with my friend Kirby, who grew up on a cattle ranch in Florida, to indulge and ask her expert opinion.

A first course of tartare, cut from the tough and hard-to-cook top round, was scrumptious mixed with egg yolk and homemade fennel capers. Kirby called it the best tartare she'd ever had in her life—high praise from a cattleman's daughter.

Following that was a plate of thinly shaved shank and then a green-onion pancake sitting on a plate next to unctuously soft chunks of beef belly and toppings like perilla leaves and fruit jam. Tucking into the loaded pancakes we'd folded into tacos, as instructed, Kirby and I were beside ourselves. It was the perfect blend of Southern textures and Asian flavors.

The entrée course was a surprise: Thinly shaved, smoke-roasted ribbons of brisket were assembled in a wreath and made for a lighter-than-bacon texture and a resounding homage to fat.

The dessert, a pea sorbet with sharp, tart herbs, cleansed the palate and had me reflecting on the irony of what we'd just eaten: This steakhouse's most prized tasting menu hadn't even featured a steak.

But whether you get a regular old New York strip or spring for the thrill-filled tasting menu, any dinner at Bateau reminds you—with the help of the ever-changing chalkboard of cuts—of its central aim of reconnecting steakhouse diners to what they're eating.

"All of this," Taylor said, "is about closing the sourcing loop."

METROPOLITAN GRILL: THE JAPANESE WAGYU BEEF YOU'VE NEVER HEARD OF (NO, NOT KOBE!)

A trip across town took me to another, very different steakhouse.

It turns out that our office literally shares a building with one of the Pacific Northwest's best steakhouses, the Metropolitan Grill.

What drew me in one blustery day was that it offered not one, but three varieties of grain-finished Wagyu, each imported from a different prefecture of Japan.

Chef Eric Hellner showed me in to the Metropolitan Grill's ornate bar off the main dining room, where we would be tasting each of the three varieties: Hokkaido Snow Beef, Ohmi Gyu, and Sanuki Wagyu.

Japanese Wagyu beef sears on a steel plate.

"What I like about them," he explained, "is that these are the three smallest, most exclusive programs—the best Japanese beef you can find in the world."

Genuine Japanese A5 Wagyu like Kobe beef is rare enough. The types of Wagyu Chef Eric offers are on a whole other level of exclusivity.

And even though USDA Prime "is still the backbone" of Metropolitan Grill, he explained, "Diners are more open to a wider variety of culinary experiences than they used to be."

Chef Eric disappeared to the kitchen and returned momentarily with a gleaming white plate filled with three tenderloins.

He pointed to each steak in turn.

"Snow, Sanuki, Ohmi."

I picked up my steak knife and cut off a piece of the steak closest to me, the Snow beef. The knife slid through it as if it were a stick of just-colder-than-room-temperature butter, something not possible with Prime-grade Angus steak.

I took a bite, chewed. The tenderness was supreme, the flavor mild—hardly beefy at all, but exceptionally subtle and complex.

"The Snow," Chef offered, taking a bite himself, "is delicate. Tender. Even slightly sweet."

Next I tried a strip of the Sanuki. The crispy outside revealed a soft, fat-filled interior—shocking for a tenderloin, which is usually the leanest part of the cow. I could see the thick streaks of intramuscular marbling intact as translucent, shiny stripes. The fat tasted stunning—of warmth, salt, and a little something that I couldn't quite place.

"The fat on the Sanuki beef is smooth." He popped a bite into his mouth. "It's incredible."

The Ohmi Gyu gave my knife more resistance when I cut off

A GUSTATORY TOUR OF JAPAN

Hokkaido Snow Beef

Hokkaido Snow Beef is perhaps the rarest beef in the world, with only one producer in all of Japan raising it.[47] Hokkaido is the name of Japan's northernmost and snowiest island, renowned for deep powder and "snowily" marbled beef. Like the other top-tier regional variants of Japanese beef, Hokkaido Snow can reach marbling levels of A4 and A5, but is much harder to track down than Kobe of equal quality rank, so it carries more mystique. At last count, there are only four restaurants in the world that serve Hokkaido Snow Beef, three of which are in the United States.[48]

Sanuki Wagyu

Like Hokkaido, Sanuki Wagyu is niche even within Japan. Sanuki beef comes from Japan's Kagawa region and is especially rich in glutamic acid (which creates a stronger umami flavor) and oleic acid, one of the much-lauded healthy fats. With only a few Sanuki cows slaughtered per month, it's some of the hardest-to-find beef in the world.[49]

Ohmi Gyu

Along with Matsuzaka and Kobe, Ohmi Gyu is one of the three most famous beef breeds within Japan. Ohmi Gyu beef comes from Shiga Prefecture and is believed to be the official Wagyu of the Imperial family.[50] Like Sanuki Wagyu—or any beef of the Kuroge Washu breed—Ohmi Gyu boasts extraordinarily high levels of monounsaturated fats because of high levels of oleic acid.[51] Its flavor is reminiscent of caramel and cigar smoke.

a small square. It was decidedly less tender—relative to other Japanese Wagyu anyway—but the flavor was much beefier, almost woody tasting, reminiscent of cigar smoke. Ohmi Gyu was like a much more refined, exquisite version of an Angus steak compared to the Hokkaido and Sanuki.

"I prefer the Ohmi," Chef declared.

When I asked why, he replied, "With the A5, or these others"— he gestured at the steaks on our plate—"they eat more like foie gras. The Ohmi Gyu, though, is not quite as rich, allowing you to eat a larger portion."

I asked Chef what he thought about the comparison between steak and wine.

"People get a little bit over the top in their descriptions," he said. But then he added, "in this Sanuki Wagyu, I will say you *do* taste a little bit of prune, and almost an olive flavor that comes through in the meat from the feed."

As I chewed, taking in another bite of the Sanuki, I realized he was exactly right. There was a salty, fruity flavor to the Sanuki that *prune* matched pretty perfectly.

The Metropolitan Grill is the quintessential craft steakhouse with its small-batch offerings, but it was far from alone in its transformation. All of the steakhouses I had visited—and many others—are pushing the envelope far beyond typical American steak offerings and developing beef menus as sophisticated as the wine list. American steakhouses are going craft.

CHAPTER 3

THE OTHER
88 PERCENT

STEAKS WORTH (RE)DISCOVERING

If you're like me, when you picture steak, you imagine a
perfectly seared tenderloin, ribeye, or New York strip—the
so-called "Hollywood cuts." But as it turns out, those familiar,
much-loved portions only amount to 12 percent of the cow,
which begs the question: What might we be missing out on?

Cuisines from around the world only increase the sense
that we Americans are wasting an opportunity for gustatory
delight. Walk into the most tantalizing-smelling restaurants

in Hanoi, Osaka, or Mexico City, and you'll find that many of the dishes reeling you in by the nose are centered on cuts of the cow entirely outside your culinary vernacular. Our situation in American steakdom amounts to nothing less than dinner-table tragedy: We've forgotten 88 percent of the cow.[1]

* *

One summer day in Seattle, I found myself in a chilly butcher shop at the corner of Queen Anne Avenue and McGraw Street. The smell that hit my nostrils as I stepped toward the counter was rich and sharp—a little meaty, a little peppery—with just a whiff of antiseptic. I was at B&E Meats, a Seattle institution dating to the 1950s with multiple outposts across the city, including this one near my apartment. I was here because Joe and Ethan had advised that it was high time I reacquaint myself with the cow—the *whole* cow.

But before I did that, I wanted to understand how and why we came to snub the vast majority of the most important meat animal on the planet.

THE PROBLEM WITH ABUNDANCE

Our ribeye monogamy can be traced to America's unique state of beef abundance. The United States is favored with vast grasslands and plentiful water, factors that make large swaths of the nation well suited for grazing animals. That's precisely why buffalo herds thrived by the millions before they were hunted to near extinction. It wasn't long after the first European settlers arrived that America became a place of "carnivorous abundance"—distinctly bovine abundance, to be specific.[2] And that's only become truer as the years have gone by.

But it's not just natural resources that made us the world's largest producer of beef, with an astounding annual production volume of 12.4 million tons.[3] No, the more decisive reasons beef became as ubiquitous and accessible a part of American life as it is were railroads and refrigeration. They created this crazy thing called boxed beef.

BOXED BEEF

Subprimals or retail-ready cuts (primarily tenderloins, ribeyes, and New York strips) are packaged in vacuum-sealed plastic at packing houses, boxed, put in the back of semis, and sent to retail grocery stores around the country.

By the 1960s, boxes packed with pre-vacuum-sealed, wholesale cuts like tenderloins or ribeyes began arriving at supermarkets[4] and restaurants by the truckload. Boxed beef was a newfangled concept in those days, and it won the pocket-books of shoppers because it was convenient and cheap, much more so than the offerings at a butcher shop sourcing its cows from a local farm.

The butcher shop itself started to change around that time too. Most butchery jobs migrated out of the neighborhood nook and into the grocery store, where the craft of cutting was reduced—to steal a great phrase from Kim Severson—to the art of slicing.[5] And as for the butcher shops that did survive the advent of boxed beef? Many clung to life because they too started selling boxed beef. It was a pragmatic decision for a small business, in a time when consumers were being conditioned to pay dirt-cheap prices for protein. But it had a serious implication: Our collective forgetting of the cow was happening not only in hyper-modern grocery-store aisles but also at the very heart of the American meat tradition—the mom-and-pop butcher shop.

The crux of the problem was this: Boxed beef kicked off a new era wherein a few cuts could become popular, and the rest of the animal could be forgotten. Beyond the popular 12 percent—the tenderloins, ribeyes, and strips—all the other cuts went *poof* and dropped right out of our culinary imagination.

The rest of the cow doesn't get thrown away, but for the most part, it doesn't make it to American kitchens. What isn't ground or shipped to grocery retailers finds its way into international markets like that of Peru, where high-end restaurants bid for American beef hearts and serve them up as entrees, or Egypt, where American beef liver is in high demand.[6] Another final destination for forgotten cow parts (and of course, the truly inedible parts) is rendering plants, where industrial grinders process everything from cow leftovers to dead zoo animals to create "meat and bone meal," which in turn becomes pet food, livestock feed, and fertilizer.[7] The medical industry claims the rest, like the nasal septum.[8] Yum.

But things are changing. Growing interest in whole-animal eating, local food, and international cuisine is amounting to a delicious and delightful rediscovery of underappreciated cuts from oxtail to tongue, even prompting a revival of the small-scale, custom slaughterhouse (a far cry from the industrial-scale ones that process up to 400 animals per hour) and the true craft butcher shop. That hot July afternoon at B&E Meats marked the beginning of my journey to eat what were then, to me, "the weird cuts" and what would become, in a few months, my favorite cuts.

A STEAK NAMED FLAP

As I peered down into the glass case at B&E Meats, the rows of red meat arranged by cut and identified by small chalk signs with neat lettering, one steak arrested my gaze—it was *huge.* Large, several shades darker than the crimson of grocery-store ribeyes, relatively free of white marbled flecks, and with visibly thick muscle grains, it was the size of an oblong throw pillow. It was exactly the kind of thing I had no idea what to do with. So naturally, I had to have it.

Its chalk-lettered sign read "Bavette," but when I inquired with Tim, the nice young man behind the counter, he told me it also goes by sirloin flap steak, or just flap meat. I decided privately that I had never heard a worse name for a piece of meat, or for anything at all, as a matter of fact. But as I discovered after some arduous Googling, the ugly nickname is just the beginning of this cut's singular nomenclature confusion.

Bavette is a French word applied with equal frequency to two different cuts that in America are called flank steak and sirloin flap steak, respectively. In France, there's *bavette d'aloyau,* which translates to "bib of the sirloin" and really means skirt steak. There's also *bavette flanchet,* or "bib flank," which corresponds to the American flank steak. Both are sometimes incorrectly called hanger steak. To make matters more confusing, the sirloin flap, which for all intents and purposes we can call the *correct* name for the cut, is sometimes called sirloin tip in New England or is sliced into strips, cubes, or thinner slabs and inevitably referred to as bavette all over again as a catch-all name for any especially thin steak.

Steak names and pseudonyms can be enough to make your head spin. It's a territory with few enforcers, because national and regional traditions of meat cutting have historically had so much overlap, evolution, and change. There may be as many as a few hundred cuts of beef on an animal, which you realize when you compile the cut plans from different cultures around the world.

But after all the hubbub and disagreement, I can say with certainty that the sirloin flap—and I mean the whole, five-pound thing, which is what I had—comes from the bottom sirloin primal. While the nearby flank steak has made something of a hipster comeback in the last couple years, its

DELICIOUS CUTS FROM AROUND
THE COW (AND WORLD)

Teres Major. A tiny muscle cut from the shoulder, the teres major (also commonly known as the petite tender, or bistro steak, or shoulder tender) is one of the most tender value cuts you can buy. It's lean, so it's best paired, Seattle Chef Brock Johnson of Dahlia Lounge recommends, "with a rich accompaniment like buttered potato or creamed spinach."[9] Its relative obscurity means it's much cheaper than its fancy cousin, the filet mignon.

Vacío. Another name for the sirloin flap steak, the vacío is a cut that is much more popular in Argentina and France[10] than in the U.S. In South America, it's commonly prepared as *asado*. In France, it makes a regular appearance as *steak-frites*.[11]

Mollejas. Typically grilled or pan-fried, the mollejas—or "sweetbreads" in English—are the thymus gland of the cow. Arturo Ramón II, a frequent guest on various food shows who is currently filming with México Travel Channel's *Grill Across America*, remembers eating sweetbreads while growing up on the Texas-Mexico border near Brownsville. "Sweetbreads are definitely high in cholesterol and fatty, but when you cook it right, it's so, so good."[12]

neighbor sirloin flap continues to fly below the mainstream radar and as a result is half the price of flank.

I got home with my butcher-paper-wrapped steak and began to realize I faced a dilemma greater than name confusion: preparation.

Karubi. One of the most popular cuts for yakiniku-style grilling in Japan, karubi is the boneless short rib.[13] It's one of those cuts that's much more marbled on a Wagyu cow than on an Angus, and it's in high demand at yakiniku joints for its mild, delicious flavor. Unlike the large, chunky short ribs you'll see in America, karubi is cut very thinly and in highly geometric shapes. If you look at a plate of beef ready for grilling at a yakiniku restaurant, you'll immediately notice that Japanese cuts involve more perfect right angles than American cuts. It's a pretty beautiful thing to behold.

Morcilla. Spanish blood sausage, morcilla can be made from pork or beef blood and is similar to what in England is called black pudding.[14] It typically tastes earthy and garners mixed reactions. Many adventurous eaters swear by its deliciousness.

I put flap—or flappy, as I had re-christened it on the walk home—still wrapped, into my cast-iron skillet for fit. It hung defiantly over the sides.

I supposed I could cut it into more manageable slices; that would make for a bunch of small steaks. But, I decided with a confident nod, that was less adventurous, and thus unworthy. I would cook the whole thing.

I preheated my oven (taking a shot in the dark and setting it to 425 degrees) and retrieved a 9-by-13-inch glass casserole dish from the drawer below the oven. Flappy fit perfectly inside it.

I pulled out salt, pepper, and—feeling slightly giddy with the *what-the-hell* attitude I'd adopted to make light of my ineptitude—coffee. I rubbed the meat thoroughly with the grainy, spicy mix, thinking that if the coffee didn't work, at least I'd get an early-evening hit of caffeine, and into the oven the casserole dish went.

Forty-five minutes of thumb twiddling later, I pulled the hulking beast out of the oven. Its red exterior had transformed into a formidable, coffee-scented crust and was swimming in an eighth-inch of brown-and-red juices. My tiny kitchen was filled with scents: beefy, caramelized, coffee, nutty, peppery.

Bearing Anthony Bourdain's relentless, insistent, *printed-in-every-outlet-on-the-internet* admonishment in mind ("to cut into a steak before letting it rest is tantamount to mortal sin!"), I gave the five-pound meal a good 15 minutes to seal in its juices, while saying a prayer in my head that it would prove edible.

The wait period observed, I pulled out the biggest plate I could find and transferred the steak to it. I stood over the counter, the smell wafting up tantalizingly, and decided to dispose of the nicety of sitting while eating.

I sliced off a strip, sawing through the large, still-defined muscle fibers that, while stringy in appearance and suggesting toughness, gave way with little fight. I popped a bite into my mouth.

My instantaneous reaction was surprise: This piece of meat contained more juice than seemed possible.

I chewed and noted that it was not at all fine-grained yet was decidedly tender, each fiber distinct but soft. What overpowered my impression of the texture, however, was the depth of flavor: It was a warm, slightly minerally, malty beef flavor that lasted through to the final chew. With each bite, I got a quick burst of caramel, and the pepper-coffee-salt added a dash of heat that made the meat itself taste sweet by contrast. I was not entirely sure, after all, that I would have leftovers.

* *

SHANK

My next steak adventure began with a lucky friendship. Ethan happened to know "Chef in the Hat," a much-adored Seattle chef named Thierry Rautureau, who hailed from rural France and helmed two French restaurants downtown called Luc and Loulay. Chef Thierry, Ethan told me, was bound to be a source of unorthodox beef inspiration.

That's how Ethan and I came to find ourselves sitting across from Chef Thierry Rautureau on a Tuesday. We were meeting for an early lunch to talk all things beef, and between indulgent bites of Loulay's signature burger—dripping with duck egg yolk, melted cheese, and foie gras—I managed to ask whether Chef had been preparing any unusual cuts recently.

Beef shanks.

He leaned toward us and said, "My new favorite thing is beef shank."[15]

"Shank" brought to my mind the image of a sharpened toothbrush, but that clearly wasn't what he was talking about. Beef shank, Chef Thierry explained, is a cross-section of the leg including a small cylinder of bone encircling an even smaller, unbelievably rich nugget of marrow.

Traditionally, Chef told us, "you would prepare a shank with a braise."

Cooking the tough cut slowly in liquid helps to tenderize it and transform the hard collagen into soft, unctuous, melt-in-your-mouth fat. A classic French braise of shank is the *daube de boeuf Provençale*, a slow-cooked stew prepared in a round clay plot. But that's not how Chef Thierry was doing it.

A few weeks prior, he had been hosting his mother at his house in Seattle.

Wanting to treat her to a simple, decadent grilled steak for their final dinner before she left town, Chef Thierry had reached deep into his freezer and groped around for whatever his usually well-stocked stores would lend him for the evening.

But to his surprise, he came up empty—of anything but a few rock-solid, frozen shanks, that is. He hadn't blocked the requisite three-plus hours to prepare shanks in the usual way, but by that point, he had few other options. So, throwing caution to the wind, he resolved to do the unthinkable: grill a tough cut of meat.

"I decided I would do them on my barbecue," he said, "just five minutes each side."

He set about preparing, sizing up the challenge of his unconventional approach, and decided that the barrier to an edible, non-braised shank was in that tough ribbon of collagenic membrane surrounding the leg. So he got creative.

"I did little slashes all around the outside of it, lots of tiny little marks with the knife," he explained.

Then he grilled it.

To the typical American palate, there's only one rule when it comes to tenderness: The more tender the better. A tough steak, we say with utter conviction, is a bad steak. And that can certainly be true, but other food cultures don't place quite the same premium on meat softness that America does. Chef Thierry was open to different textures and firmness.

"We cut into it, and it had a little chewiness to it—a good chewiness—and it was so, so, so good," he relished, eyes closing at the memory.

"My mother loved it. It was frikking delicious."

Chef handed over that recipe all too willingly ("One day we will barbecue it together!"), and I took it home, remembering the critical cross-hatching step, to try it for myself.

THIERRY RAUTUREAU'S RECIPE
FOR GRILLED BEEF SHANKS[16]

Servings: 4–5

Ingredients

- 4–5 beef shanks (about one pound per person; each cross-cut shank will be approximately one pound)
- Dry rub of your choice
- 2 Tbs. olive oil

Preparation

- Defrost frozen shank, still wrapped in plastic, for three hours or as long as needed in a bowl of cool water.
- Using a sharp knife and holding the shank carefully, slice through the hard, rubbery membrane surrounding the cylindrical shank. Thierry advises: "This is the trick of the grilled shank. You *must* take your knife and cut through it in little slashes, make little marks all around the outside of it."
- Season with your favorite dry rub, and let it sit for about an hour on the counter.
- Next, preheat your grill to high. Once extremely hot, move the coal to the side so you have a space of indirect heat.
- Just before placing the shanks on the indirect heat side of the grill, "brush on top of the shanks a teensy bit of olive oil."
- Place shanks on grill and cook approximately five minutes on each side.
- Remove shanks from grill, let rest five minutes. Then slice and serve.

I used shanks from Hutterian Farm, a self-sufficient farming community of Anabaptists in the Southwestern corner of

Three cattlemen at Hutterian Farm in Reardan, Washington. The Hutterites are a self-sufficient Anabaptist community who farm for a living: primarily no-till wheat and grain-finished, soil-carbon-friendly beef. Ed Gross, head cattleman, is center.

Washington State, because they happened to be my current favorite for pasture-raised, grain-finished beef. Unlike industrial-scale feedlots, their Black Angus grain-fed cows lived their entire lives on the same farm and received a feed of homegrown corn and peas raised in untilled soil to protect soil carbon stores.

The flavor of Hutterian New York strip, which I'd already tried, was the most delicate, soft, elevated steak flavor I'd ever encountered. It was the definition of easy eating. The shanks, by contrast, I expected to be tougher, rougher—and I didn't know what else.

I did just as Chef Thierry said, cutting careful but determined slices through the surprisingly tough outer membrane. But I diverged from his preparation in one important way: I had no grill, so I opted for my cast-iron pan. That didn't worry me. If anything, the pan would do a better job of catching errant juices.

Adding in a few cloves of skin-on garlic, I placed the shanks, rubbed simply with salt, in the extremely hot pan.

I cooked them just three and a half minutes each side, a degree or two shy of medium rare, worried I would overcook the already-tough meat.

After letting them rest a few minutes, the anticipation was killing me. I cut off a small triangle of one shank—it was definitely somewhat difficult cutting—and brought it to my nose. It was a richer smell than the Hutterian New York strip steak had been. I took a bite.

It had an extremely deep beefy flavor and was chewy compared to a New York strip. But the chewiness was pleasant, challenging but not bothersome, because the flavor—much stronger than the other Hutterian cuts I'd eaten—lasted.

The shank really tasted, and felt like, I was eating a totally different part of the animal, a working muscle with different responsibilities and, as a result, a completely different taste profile. I felt a little bit the adventurer eating this cut in such an uncouth way.

IS GRAIN OKAY FOR COWS TO EAT?
THE ANSWER IS A RESOUNDING: 'IT DEPENDS.'

The question of whether grain is healthy for cows to eat has divided foodie and farmer communities alike for years. But like most aspects of caring for animal livestock, the answer is a resounding, "It depends on the management." That's good news for all the fans of grain-finished beef out there—it means a conscientious farmer can raise delicious grain-finished beef from cows that lived healthy, stress-free lives.

Why the debate? Well ruminant digestive systems are designed for grass.[17] But those funny extra-stomach organs, which enable cows and sheep—unlike humans—to put the energy locked in fibrous stalks to use, are built with a fair amount of wiggle room[18] for other energy sources like cereals, fruits, legumes, and vegetables. It makes sense: Grains are simply the fruiting bodies of grasses; so in the course of foraging, some grains are inevitably consumed, and have been since the beginning of cow time (though before the dawn of agriculture, grains were much smaller than they are today). Prehistoric aurochs, the hulking, rhinoceros-sized beasts from which today's much smaller cattle are descended, grazed their way across continental Europe on whatever they encountered: grass in the main, but also some clovers, brassicas, grain forages, or fallen fruit.[19] Every cow's diet should consist mainly of forage or roughage, but a little grain for the sake of flavor doesn't harm them if administered right.

Grain—particularly processed corn, soy, and distillers grains[20, 21]—has become a popular cattle feed in the United States because it's cheap,[22] putting weight on a cow faster and more cost effectively than any other option, *and* because it produces a predictable, mellow flavor Americans have become accustomed to. Cost cutting and product uniformity are the exact reasons why big beef is built on grain-finished, rather than grass-finished, cattle.

The risks to cattle health begin when concentrated feedlots get *too* focused on squeezing out costs and fattening cattle at lightning speed. "Pushing too hard" with a too-intense ratio of grain to roughage, says animal welfare expert Temple Grandin, increases the risk of acidosis, a condition that throws the cow's rumen out of whack and causes all sorts of conditions from listlessness to liver abscesses to death.[23, 24] Many feedlot diets are comprised of up to 80 or 90 percent corn, which is inarguably too high.[25] Some feedlots try to solve for the danger of overfeeding grain by adding preventative levels of anti-bloat antibiotics to the daily feed ration, which gives them license to increase the grain-to-roughage ratio even higher.[26]

But grain-finished beef needn't be that way. Independent farmers in tune with their herd can design feeds that foreground forage and include some grains for mellower flavor and marbling at amounts that are fine for cattle health. Some farmers even use unusual grains like millet or fermented barley left over from the beer-production process to create more complex flavors that corn-fed feedlot beef simply can't develop.

In general, people dramatically oversimplify the equation for cattle health. It's not just *open pasture + blue skies = happy cows*. It comes down to management.

Take grass-finished beef, which many foodies assume is a silver bullet for cow health and happiness. Though legumes like clover and alfalfa get lumped into the category of "grass" for the purposes of marketing, they're actually different than grass. Eating too much of certain legumes can cause a similar condition to grain overfeeding called pasture bloat.[27] A good grass-finished farmer is constantly monitoring his pastures for the ratio of grass to legume, even though to any activist or grass-fed certifier who walks onto the property, the green stuff all looks the same. You can call any cow standing in a field "grass-fed"—including cows in overcrowded feedlots being fed from "grass pellets"[28]—but it takes a farmer with a deep understanding of land and animals to create *both* healthy cattle and delicious beef.

Whether you're talking about grain-finished or grass-finished beef, the health of the cows has less to do with precisely what they're eating and much more to do with the management style of the farmer. That is not, however, a license to feed cows candy bars.[29]

After several more delicious bites, my jaw slightly tired, I broke out a bottle of $8 Beaujolais wine to finish out the meal. I resolved I would one day try the traditional French method: the *daube*.

TONGUE

If you asked me a year ago what cut of beef I would absolutely never eat, my ready-made answer would have been tongue. So it followed that once I decided I may as well do the thing properly and get to know the whole cow, I had to face the tongue.

Something about the visible taste buds on a whole, two-pound cow tongue posed a not-insubstantial psychological challenge. Yet I knew it to be a celebrated cut in other parts of the world; lengua tacos were already spreading like wildfire across the southern half of the U.S.

When I spoke with *Texas Monthly*'s barbecue editor, Daniel Vaughn (who holds the first position of its kind in the world and possibly also the most coveted job anywhere), he told me that he thought the American barbecue world, too, was about to pick up on the unusual cut.

"Tongue," he said, "is ripe for popularity. It's basically like a miniature fatty brisket."[30]

I wanted to give tongue the true lengua treatment, so I called up someone for whom tongue's deliciousness was no secret: Mely Martinez,[31] the food blogger behind the much-loved blog *Mexico in My Kitchen.*

Mexican food culture, Mely told me from the other end of the line one morning as I sat looking out my window at the Seattle drizzle, celebrates the tongue above most other cuts.

The texture is, by her account, nearly indistinguishable from ground beef when you chop it finely, but the taste is far more flavorful.

She explained, "There is a saying in Mexico that goes, 'The two ends of the cow are the most delicious parts.'"

She paused for half a beat and then added, "That would be the oxtail and the tongue." I supposed she was protecting against the possibility that I might not know what the "ends" mean. She wasn't wrong.

Mely's blog began as a digital cookbook of Mexican cuisine for her son, who she hoped would stay closely connected to his culinary heritage despite growing up in the Northeastern U.S. But over the years, *Mexico in My Kitchen* grew to touch the lives of more people than just Mely's family. Flocking to her website in droves were second-generation immigrants seeking to recreate dimly remembered dishes, the American spouses of Mexican-Americans wanting to learn how to cook a comfort food, and adventurous Americans unfamiliar with Mexican cuisine but introduced to it thanks to the power of the internet.

As she listed cuts popular in traditional Mexican cuisine— organs, oxtail, and tongue chief among them—Mely reminded me of an important point:

It's not just that America is ignoring *edible* portions of the cow; we're ignoring parts that other cultures consider the pinnacles of gustatory delight.

The cuisines that claim some of the most exquisite beef dishes (Vietnam and its pho, South Korea and its kalbi barbecue ribs, or Peru and its beef-heart kebabs) typically mastered the preparation of those cuts out of necessity. When calories are a scant resource, to waste the shanks, short ribs, or heart after getting through the easier-to-prepare cuts isn't an option. Over the course of generations, preparation methods and recipes were honed to turn those more difficult cuts into not just edible and nutritious dinners but enviable delights to make even the ribeye-spoiled swoon.

Mely gave me her tongue-tacos recipe, and I took it to my kitchen with relish and only minor trepidation on the afternoon that I managed to procure a whole, hulking cow tongue—this one from Gebbers Cattle Company in Central Washington State, where the Angus cows are raised on sparsely forested, mountainous rangeland in the shadow of the Cascades.

Perhaps it's something about their feed (a grain mix with apples and pears from the Gebbers's orchard operation), or perhaps the climate (dry, either hot or very cold, and unsuited for grass but perfect for corn and alfalfa) made the flavor of Gebbers beef significantly stronger and sharper than Hutterian beef.

It had surprised me how I could pick up those nuances after only just a month or so of eating steak regularly.

I grimaced as I weighed the tongue on my digital kitchen scale: It clocked in at 3.3 pounds.

Corina Gebbers, a fifth-generation rancher at Gebbers Cattle Company, helps her dad and brother on a summer cattle drive in Brewster, Washington. The Gebberses partner with state and federal agencies to conserve their expansive forest and rangelands east of the Cascade Mountain Range.

It was exceptionally . . . tongue-like, but I tried to embrace that thought—its *tongueness*—as I brought a pot of salted water to a boil, added a bay leaf, and remembering Mely's instructions, placed the entire tongue into the boiling water. I reduced it to a simmer and covered the pot tightly.

Three hours later, I found myself with a rather large, somewhat knobbly, but definitely thoroughly cooked cow tongue. It was now gray. Not the most appetizing-looking thing I'd ever seen in my life.

Black Angus cows graze in the forest above Brewster, Washington, on a misty August morning.

Taking it out of the pot, I placed it on a plastic cutting board, where it steamed gently. The next step was going to be the grossest.

I pulled opened a drawer, considered the handheld vegetable peeler lying innocently on the drawer liner. No. Too nasty. I would use a paring knife for this. I had to remove the taste buds and the thick skin that covered the tongue.

I gritted my teeth and wielded my knife.

I found, after 30 seconds or so, that the removal of the taste buds sort of stopped bothering me. I still felt hyperaware that

I was handling the tongue of a previously living creature, but somehow, inexplicably, that fact just stopped being offensive after a while. I felt kind of . . . cool. The cow this tongue had once been attached to was being made full use of. Nothing wasted. Everything—even the parts that look a little intimidating at first—honored. Made delicious use of rather than discarded or pawned off to the rendering plant. It also smelled good by this point, which helped.

MELY'S RECIPE FOR TACOS DE BARBACOA DE LENGUA DE RES ("BEEF-TONGUE TACOS")[34]

Includes Mely's notes!

Servings: 4 servings (per pound of tongue used)

Recipe Notes: Beef tongue is still very affordable; buy it in a place that you trust for their quality. The way I serve it at home is buffet style with the tortillas, chopped onion, cilantro, salsa, some lime wedges, and salt. Let everyone prepare their own tacos *¡Buen provecho!*

Ingredients

- 1 beef tongue
- 1/4 of medium-size onion
- 4 garlic cloves
- 1 bay leaf
- Salt to taste
- Water enough to cover the beef tongue

The taste buds peeled off, I sliced the tongue into thin strips and then chopped it into eighth-inch bits until I had something that resembled a slightly more cubical than usual version of ground beef. My mind traveled briefly to the ground beef you can get from the grocery store, a single package of which comes from as many as 100 cows.[32] This single-tongue version of taco meat was in that way less bizarre than 'regular' ground beef.

To serve

- Warm corn tortillas
- 1 cup of chopped cilantro
- 1 cup of chopped white onion
- Salsa of your choice

Instructions

1. Rinse the beef tongue with water and place in your slow cooker. Add onion, garlic, salt, and water. Cover and set on low for eight hours. Cook until tender. If after eight hours the meat is not tender enough to shred, cook a little bit longer. Not all slow cookers work alike.
2. Remove the beef tongue from the crock pot and place in a large dish. Remove skin using a knife to make a cut at first and discard. Trim off any fatty tissue at the bottom end.
3. Shred the meat using two forks and place in a serving bowl with some of the cooking broth that has already been degreased and strained.

My plans were to house the chopped tongue in corn tortillas, salsa verde, and avocado, but first, I couldn't resist trying a tiny cube of the tongue on its own.

I chewed, and it was . . . delicious. With a way much percentage of fat calories than most muscle meats,[33] ground tongue was like ground beef on steroids. Seriously premium taco filling.

I pulled an avocado out of my roommate's avocado stash and cut a slice. Popping it into my mouth on the same bite as another cube of beef, I got the creamy fat of the avocado, the beefy umami of the meat—wow. Would I even get to tacos?

MERLOT STEAK: A FINAL RECOMMENDATION FROM A BEEF SOMMELIER

It's much more important to know your butcher than your stock broker. Playing the stock market is a fool's game, and I do it. Sometimes you win and sometimes you lose. But knowing your butcher has real benefits. Unless I'm just going for something ordinary, I almost always ring the bell at the butcher counter.[35]

—MEATHEAD GOLDWYN

My quest to discover weird cuts had taken me to my neighborhood butcher shop, a French restaurant, and a Mexican food blog. *New York Times* best-selling author and grilling expert Meathead Goldwyn—who had adopted the moniker a

A TURNING POINT FOR CRAFT BUTCHERY

In 2004, when restaurateur Danny Meyer (of Shake Shack fame) asked third-generation New York meat purveyor Pat LaFrieda to create a custom blend of hamburger meat for his Shake Shack burgers,[38] the visibility helped launch LaFrieda, 34 at the time, to the national foodie stage, and with him the mystique of the skilled neighborhood butcher.[39]

Now, over a decade later, places like LaFrieda Meat Purveyors and Fleishers Craft Butchery receive countless requests from aspiring apprentices and interns willing to work for free to learn the old trade,[40] a sign that the craft butcher shop is on its way back. And indeed, many of the figures in the butchery renaissance are of the younger generation who, in the absence of formal educational opportunities in the U.S., taught themselves with books or cobbled-together internships.

journalist once gave him, "barbecue whisperer, hedonism evangelist"—told me there was no one more important to consult for meat than the guy behind the butcher counter, so I decided I should round out my education with the help of a true craft butcher. Luckily I knew of such an expert through a friend, so one October morning when I was driving through Pennsylvania farmland, I called up Brad McCarley[36] in Memphis.

In the past 10 years, Brad told me, good things had been happening for small-scale, independent butcheries—America was going through what the *New York Times* called a butchery "renaissance."[37]

Brad broke into butchery when he stumbled across a Craigslist ad that asked: "Want to learn the art of butchery?"

For $11.50 an hour, Brad quickly fell in love with European-style butchery, which he described as being less about "hand saws, as in the American tradition, than about knowing how to follow the seams."

He explained that even in the European way, knives weren't completely removed from the equation: "Maybe a knick here, an incision there—but from then on out, you're really just pulling muscles apart."

He eventually landed a job at a Memphis grocery store called The Curb Market that sourced all its wares within 60 miles and prided itself on having an artisan in-house butcher shop sourcing beef from the proprietor's own farm.

He sees a need for butchers to give Americans a culinary re-education. "People tend to want only 12 percent of the animal—tenderloins, strips, and ribeyes," he explained.

"That's what makes local beef, and the whole-animal butcher thing, really difficult. There are more great steaks on a cow than most Americans realize."

Brad has been taking it upon himself to tell his customers

A merlot steak.

about lesser-known cuts, and slowly but surely, he's seeing a change in their willingness to experiment beyond the typical griller's fare.

"We spend a lot of time talking to customers who haven't heard of those other great cuts, because it's just not what's generally sold. We explain the anatomy of the animal and explain where they would find the flavorful cuts most people don't know about."

One of the least-appreciated cuts on the cow is the merlot steak.

"It's one of those cuts that someone who came up in the American supermarket tradition of butchery might not know about."

It comes from the backside of the lower rear leg and is known for its velvety texture, comparable to the much more expensive tenderloin.

"It's really flavorful too," Brad added, "but it's really hard to find unless you know exactly what you're looking for."

His customers are convinced—once they try it, they come back wanting more. "But the problem," he said, "is that there are only two merlot steaks on each animal."

* *

I found one of those merlots for myself—a grass-finished merlot steak from Harlow Cattle Company, an hour's drive south of Seattle—and seared it to a quick medium rare the next evening for dinner.

Though grass finished, which is typically known for being tougher than grain finished, it was as velvety as any tenderloin I'd ever eaten. The texture, if I were to sum it up, was like slicing through a block of very cold cream cheese. The flavor, however, was much richer than the notoriously taste-lacking tenderloin.

The merlot was a tenderloin in Technicolor.

CHAPTER 4

A SECOND CHANCE FOR SMALL FARMS

———

CRAFT BEEF IS GOOD NEWS FOR INDEPENDENT RANCHERS

———

Two dozen large black cows were lumbering through a maze of fencing in the early-morning sunshine, on their way toward a rusted scale that Becky Harlow Weed[1] was using to weigh each Angus-Hereford steer. On this cold spring morning we (Joe and Ethan) were at Harlow ranch with Caroline outside Spanaway, Washington, where the snow-covered peak of Mount Rainier rose up 50 miles beyond the barn. The mountain looked close enough to touch.

The view of Mount Rainier from the pastures of Harlow Cattle Company on the Camas Prairie.

Unlike her father, who had ranched before her, Becky was raising her cattle all the way from birth until slaughter. Previously Harlow Cattle Company had been a cow-calf operation, selling weaned calves at auction, where they were bought and transferred to a feedlot, processed at an industrial-scale slaughterhouse a few months later, and their most popular cuts finally distributed through the commodity beef system. By raising grass-fed and grass-finished cattle entirely on pasture, Becky was bucking the standard industry practice and charting an independent course. And she had done so for a very simple reason.

"It was pretty much impossible to stay in business," she said as

Becky Weed and her dog Tipper checking on the cattle on a foggy Washington morning.

she swung a leg over the fence and nimbly dropped into the enclosure to coax forward three animals that had stopped to peer at us curiously.

"It was like I was paying them to take my meat."

Becky's struggle to make ends meet in the commodity system is the rule rather than the exception—the rise of big beef has hurt no one more than the small-scale American farmers who supply it. In the 1930s, the United States was home to 6.3 million farms. Today, that number is closer to 2.2 million and shrinking every day.[2]

Tipper surveys his kingdom.

The industrial-beef supply chain is shaped like a funnel. At the wide mouth, there are more than 700,000 cow-calf operations, 91 percent of which have fewer than 100 head of cattle[3]—by any measure, they are small farms, many handed down through the same families for generations.

Cow-calf operations represent the relatively pastoral first six months of the industrial beef-cow's life, but that's the last point on the supply chain at which anything can be called small. The feedlots—the next stage for the cow—are huge. Forty percent of all the cattle on feedlots in the United States are found at stockyards with more than 32,000 other bovine compatriots; some have capacity for more than 100,000 animals.[4] Feedlots span mile after dusty mile, reaching as far as the eye can see toward the horizon in states like Texas, Nebraska, Kansas, Colorado, and California—the five biggest states for cattle feeding in the U.S.[5] And when the cattle are moved to slaughter, it's often to processing facilities that can kill thousands of animals per day. Four meatpackers control virtually all of those slaughterhouses, and in general, the industry is tending toward higher concentration: fewer and fewer feedlots with larger and larger capacities.[6]

The American steak machine is designed to maximize efficiency at any cost and is dominated by the downstream players: feedyards, processors, and retailers. Cow-calf ranchers are left with little bargaining power.

It was after her father's death that Becky took over Camas Prairie Ranch and Harlow Cattle Company and—at the urging of her son, whom she describes as a "foodie"—began to look for new markets that might be interested in buying her grass-fed and finished beef.

"It was an opportunity to operate an actually viable agricultural business," she explained.

But it wasn't a walk in the park.

"My husband, Mark, and I have talked to Whole Foods three or four times," she explained. "And PCC," she added, referring to a popular Pacific Northwest coop.

"What you have to understand is that they don't want to work with small farms—they don't want that many individual accounts."

And even if they were willing to put up with the complex invoicing that would entail, major retailers won't pay the prices that small farmers require to stay in business; which means—in Becky's opinion—that the farmers are basically subsidizing customers' dinners.

"It's like the cattle company throwing a 20-dollar bill on the table when you order a steak," Becky laughed.

An Angus-Hereford steer poses for the camera at Harlow Cattle Company in Spanaway, Washington, an hour south of Seattle.

She was right. Traditional retailers see a logical business case for buying only from the biggest meatpackers rather than directly from small farms: simplicity of invoicing, low prices, and a consistent and predictable product. So Becky and her husband looked beyond traditional grocery stores—even beyond the ones with a reputation for being crunchy or "foodie"—and eventually found craft beef e-retailers and restaurants whose mission matched her product.

Her very first customer, a popular Bainbridge Island watering hole called Harbour Public House, gave Becky confidence and a stream of income that would soon be complemented by other restaurants and craft e-retailers as interest in craft beef picked up traction.

"Jeff McClelland, the chef there, and the owners, Jeff and Jocelyn Waite, have been great," Becky said. "They're really committed to using the whole animal."

That commitment is critical for small farmers, but it's one that most grocery retailers aren't willing to make. Ranchers face the unavoidable reality from which most American home cooks are distanced: Every part of the cow has value and needs to be sold—not just the ribeyes and tenderloins.

Since the rise of boxed beef in the 1970s, groceries have typically been interested in only the most popular cuts of beef. Big beef has developed the means by which to farm out the remaining, less popular cuts to a vast network of overseas markets, dog-food companies, and medical industries. But for a solo operator like Becky, marketing at that scale and complexity would be impossible. That's why the craft-beef markets willing to buy whole animals—e-retailers, restaurants, and even a smattering of craft butcher shops—are so important for small farms. They represent an unusual potential for viability in an industry that's notoriously unforgiving for small-scale producers.

And those emerging markets, unlike traditional grocery stores and restaurants, appreciate the quality difference between CAFO beef and craft beef.

We got a sense of the huge investment of time and effort Becky makes in her farm as we peered at the ranch notebook lying open on the workbench to our left. As Becky corralled the cattle behind us, we took in the columns of data that tracked each animal's monthly weight gain—just one small hint at the labor involved in producing high-quality, grass-finished beef.

Most days, Becky wears a dozen hats. In the morning, she might feed the horses, zip out in her four-wheeler to move the cows, repair the sections of fence that need repairing, notice a newborn calf and race back to the house to pick up materials to vaccinate and ear-tag it, load hay into the barn, check on the cows again, hand-pull invasive plants from the creek, build stream buffers, and check on the cows again for good measure. And that's on an easy day.

The added effort is worth it though, because Becky's oversight and involvement in every stage of her cattle's lives (even the slaughter, since she uses a mobile unit lent by the Pierce Conservation District—one of only a few in the country) lets her ensure the quality of the meat from start to finish.

Increasingly, craft markets are appreciating the work of farmers like Becky. It's the difference between a steak you can't trace back to a single country—much less a single farm—and a steak raised with care, at a known location, with known ingredients, and by a rancher who cares about both the life of the animal and the deliciousness of the food it becomes. Big beef has historically shied away from traceability, perhaps because they didn't want consumers to be turned off by what they saw: meat imported from thousands of miles away, raised on a crowded feedlot, and slaughtered alongside 400 other animals per hour.[7, 8] Craft beef is raised by farmers who take pride in the origin and life of the steak they raise and would happily invite you in for a tour.

As the industrial-beef chain squeezes the ranchers that supply it, it also strains the social and economic fabric of once-vibrant rural farming communities. You can describe the impact of beef's industrialization on those communities as a ghost-town effect: With much of the value of the cows going, in the words of grass-finished rancher Robert Boyce, "disproportionately to the people downstream," profits and business alike leach out of farmers' communities.

"Uncle Bob" Boyce of Lil' Ponderosa ranch in Central Pennsylvania is a farmer who not only has managed to make a living selling to craft-beef markets but has also carved out a way to keep the value created by his farm business within his Central Pennsylvania community.

Top: Uncle Bob Boyce stands outside his neighborhood meat shop in Chambersburg, Pennsylvania. It's in the same building complex as his small-scale slaughterhouse, and a stone's throw from Lil' Ponderosa farm (**above**).

AUNT KATE'S RECIPE FOR WORLD'S TASTIEST MOCK TENDERS[9]

When we visited Uncle Bob and Kate Boyce at their ranch in Lower Frankford Township, Pennsylvania, Kate prepared a scrumptious lunch of beef from their own purebred Angus herd.

We sat at a dining table laden with good food: a tureen of silky mashed potatoes, a large green salad, a platter of mock tenders, and a small jug containing a thin gravy of their pan juices.

Mock tender is a cut that causes butchers who sell it considerable distress. The main problem is that it looks exactly like a tenderloin—hence the name. But it actually comes from the chuck primal and is a muscle that is used intensively. As a result, it's very flavorful but exceptionally tough if you were to make the fatal error of preparing it like a tenderloin.

Do anything but braise it low and slow, and you'll have a hockey puck on your hands. But do it right? You'll be in for a serious treat, as we discovered when we sampled Kate's mock tenders. The flavor was incredibly rich, deeply beefy, and tangy.

Kate was kind enough to share her recipe.

For over 20 years, Uncle Bob had grown a freezer-beef business, selling directly to clients—both local and from surrounding states—and handling all the marketing and distribution himself.

But it made a major difference when he was approached by a local chef named Sean Cavanaugh, who had recently opened the Lancaster steakhouse John J. Jeffries. Chef Cavanaugh was

Ingredients

- 3 pounds mock tenders, sliced ¾ inch thick
- 2 medium-sized onions
- 2 to 3 cloves garlic, to taste
- Salt and pepper, to taste

Instructions

- In a skillet, braise meat on both sides.
- Then add water to come up to half the height of the cut.
- Add salt and pepper, to taste. Add a couple of bay leaves.
- Chop two medium-sized onions and shred over top of meat and into the water.
- Add two or three cloves of garlic, finely chopped.
- Cover with a tight-fitting lid and simmer for 2 to 2.5 hours. Do not boil. You are trying to capture the flavor of the meat.
- Serve!

looking for a dependable supplier of high-quality, grass-finished beef, and he found that in Uncle Bob. Chef Cavanaugh and his co-chef, Michael Carson, liked Lil' Ponderosa so much that they came down to the farm often and eventually became partners in Lil' Ponderosa Enterprises, holding partial ownership in the farm itself as well as the small-scale slaughterhouse they purchased jointly with Uncle Bob in 2016.

When we asked Uncle Bob what had compelled him to buy a local slaughterhouse (one with capacity to process just one or two dozen animals per week, compared to the hundreds per hour at large-scale slaughterhouses), he explained that he and Kate wanted to offer "high-quality, local beef"—what he calls "real food," processed and slaughtered under "the best possible conditions."

With the slaughterhouse a 30-minute drive from the farm and the restaurant only an hour away, Uncle Bob's farm was benefiting people throughout Central Pennsylvania, in the boost the farm gave the local economy plus the delicious steaks it produced.

"My wife, Kate, and I always ask ourselves," Uncle Bob said as we walked through a field where some of his 200 purebred Black Angus were grazing on a sunny September day, "are we doing what God would do if we weren't here? That's our test. If the answer is yes, we're doing okay."

With more and more consumers demanding delicious, ethical, and independently produced beef, it's no longer the case that commodity auctions are farmers' only option. That's not to say the problem is solved—there are more craft farms out there than the relatively small market currently knows what to do with. But as the movement continues to grow, the market for craft beef will too.

CHAPTER 5

UNSUNG ENVIRONMENTAL HEROES

—

THE ETHIC OF CARE AT CRAFT BEEF FARMS

—

With four Border Collies piled in the back of the ATV and the three of us squished in the front, George Lake, his wife Christy, and I were flying down a dirt path on our way to visit the cows at Thistle Creek Farms. It was a steamy September morning in Pennsylvania, and George was narrating for my benefit, while behind us the herd dogs panted, hopeful they'd soon be called into action.

"That barn," George said over the roar of the engine, pointing to his left at a weathered building alongside a winding two-lane road, "was a mule barn for the old flour mill around here."[1]

Christy added, "It was built way back in 1796."

We wove between the Lakes' fields of cowpeas and sorghum Sudangrass—the former a nitrogen-fixing legume, the latter a heat-loving annual full of sugar—and finally down through a forest on a second path that Christy told me had been Pennsylvania's main wagon turnpike in the 1700s.

"It went all the way across the state," she said.

We rolled to a halt next to a long expanse of wooden fence, out of which a plastic line extended perpendicularly down toward a creek far below. The bright-white span of rope was one of the many moveable fence lines that enabled George to rotate his cows to a new stretch of pasture multiple times each day.

"You're about to watch him move the cattle," Christy murmured as George leapt lightly from the driver's seat, not looking remotely his 71 years, and walked forward to kick off the day's first exercise in topsoil regeneration.

The farm dogs ready for the workday at Thistle Creek Farms in Pennsylvania.

The fact is, cattle ranching in America gets a bad rap, and is hardly known as a pillar of environmental stewardship.

Most of the burgers that end up on our plates in America come from penned animals that lived a large part of their lives on concentrated, industrial-scale feedlots. Those dusty, muddy yards, stretching for miles across Kansas and Oklahoma and Colorado, hit your nose and discolor the sky long before they come into view. The smell and the appearance are indicative of a major problem on feedlots: Nutrients can't cycle.

When cows graze on pasture, manure is a fertilizer. But when it stagnates—whether in holding ponds or as an ankle-deep layer in the feedlots—it becomes a toxic pollutant that harms surface water, groundwater, and air quality, while also releasing methane into the atmosphere. And to add insult to injury, by doing away with pasture, crowded feedlots eliminate a vital carbon storehouse that would otherwise absorb emissions: grass.

But this isn't the full story of beef production in America. Many small-scale ranchers have a strong, albeit totally un-showy, brand of environmentalism. It's in the economic interest of multi-generational farms to keep their land, soil, and water resources productive over the long-term. Increasingly, environmental groups are applauding farms for their conservation work—farmers just call that good-sense management.

George Lake told Joe, Ethan and me that for 30 years, his management focus at Thistle Creek had been on soil health, because when he inherited the land he could tell it was in a sorry state. He had returned to the farm from his post as a fighter pilot to realize his dad's chemical applications on the feed pastures had depleted the soil. How did he know? There were no earthworms— and he knew enough about Pennsylvania land to know that meant his soil wasn't living. How, he wondered, was he supposed to feed animals from plants grown in dead soil?

It was a difficult time. George had just lost a family member,

George Lake looks out over his pastures, whistle in mouth. He uses it to guide his highly trained herd dogs as they move his cattle and sheep.

and had sold his dad's dairy herd to pay the inheritance taxes to keep the farm. Keeping the land—an heirloom in and of itself—was the bottom line as far as he was concerned. But so was farming frugally. No expensive chemicals or heavy equipment, just the resources the hills could provide. He would know he'd been successful when the earthworms returned. George bought a few beef cattle and started building up a herd, while also accepting a three-day-a-week job as a commercial pilot to pay the bills. (Those cattle wouldn't bring returns for a while.)

George's frequent transatlantic flights left him with free weekends in Europe, which he spent visiting pasture-based ranches from Scotland to Romania and learning from the "grass farmers" he met there. He took the lessons back to Central

Pennsylvania and decided to implement intensive planned grazing at Thistle Creek.

Planned grazing—popularized in part by advocacy groups like Savory Institute—involves alternating periods of grazing

A farm road through the pastures at Thistle Creek Farms.

and pasture recovery. Its emphasis on pasture regeneration carries one of the most fascinating and complex benefits of well-managed ranches: carbon sequestration.

Grasslands cover 30 percent of North America,[2] and are as important a carbon sink as forests.[3] But U.S. grasslands have had their soil carbon stores depleted 30-50% due to development, livestock overgrazing, and monocropping.[4] Ranchers consciously planning around and monitoring the health of their pastures may actually represent a solution. Well-managed grazing not only leaves sequestered carbon intact, it also puts carbon back underground, or so initial research by Jason Rowntree of Michigan State University seems to suggest. Smart grazing practices may be able to roll back the carbon clock.[5]

The climate-and-beef question gets murkier when you press on with inquiries like: Exactly how much rogue carbon can be sequestered? And for how long? Can cattle farms stay carbon-negative forever? There aren't yet clear answers to these questions. We're still at the earliest stages of understanding the long-term relationship between soil and climate. (Stay tuned for more publications by Jason Rowntree.)

But carbon sequestration is by no means the only benefit well-managed farms can provide. There's value to restoring damaged ecosystems for future human use; and at Thistle Creek, planned grazing did just that. In addition, grazing ruminants—who don't mind thin, rocky soil unsuitable for row crops—can put otherwise marginal land to use for food production.

The word 'marginal' was an overly generous descriptor of Thistle Creek Farms at the outset, and George's father's chemical applications were only part of the problem. Long ago, a portion of the property had also ago been mined as a sand quarry, leaving the dirt in George Lake's pastures a drab cake-yellow color, depleted of any evidence of life. But that's where he knew cows could help.

"The really, really dark soil we get," George explained as he picked up a clod of earth, "is the result of hoof pressure." 250,000 pounds of it per acre, by his measure. The heavy bovine hooves of his herd are able to pierce the earth, pushing organic matter deep down into the topsoil and creating more humus than you can create in a livestock-free farm. While an old rule tells farmers to expect about an inch of topsoil accumulation per century, George is seeing way more.

With the grazing George's herd is doing, he explains, "we're adding an inch of topsoil every decade. That's why you need animals, and not just any animals—you need hooves."

The topsoil regeneration George is seeing isn't a perfect proxy for carbon sequestration, but the presence of earthworms *in* that topsoil is a reliable indicator of soil-carbon storage.[6] From George's perspective, the important thing is that he has restored a landscape and can count on its continued productivity. He knows that because he hit his goal. "We have so many earthworms," he said, "the soil scientists who visit us are just plain giddy."

It's not just at Thistle Creek Farms that cattle ranchers are engaging in under-the-radar environmentalism. Craft-beef

The barn at sunrise at Harlow Cattle Company.

producers across the U.S. are at the frontlines of the battle against commercial development of critical ecosystems, like Becky Weed at Harlow Cattle Company. That's what Ethan and Joe discovered when they visited Becky's farm for the first time.

* *

"I think farmers get a bad rap," she told us as we walked through a grove of Oregon oaks, a few hundred feet from the nearest cluster of her grass-finished Angus-Hereford cows.

"Most everyone who's been in agriculture for more than a few generations," she explained, "has only been successful because they are good stewards of the land."

People who haven't grown up in farming communities, including environmental activists, tend to be surprised when they learn that cattle ranchers are good environmentalists and have been for generations. It's just that ranchers don't put the "environmentalist" label on their daily work.

"The reaction when people realize what we're doing," Becky explained, "is always, 'Wow, that is so cool!'"

Becky's father had worked the Southwestern Washington prairie before her, though he operated a cow-calf ranch, selling his calves to feedyard buyers at the commodity auction because that was the only market. But when he died and Becky took over Harlow Cattle Company, she began transitioning to raising grass-fed cattle from start to finish.

At around the same time, the Pierce County Conservation District had built a mobile meat-processing unit offering USDA slaughter to local farmers, thereby eliminating the gap in Becky's equation that USDA processing had represented. The rest of her grass-finished beef business, she said, is history. But what hasn't changed from the time Harlow ranch was cow-calf to its new birth-to-harvest model was the dedication to keeping a vibrant ecosystem alive.

"It just makes sense," Becky explained, "to keep your land healthy so your cattle have everything they need."

Walking around Harlow Cattle Company, evidence of Becky's inclusive approach to farming abounds. Raptors like eagles and red-tailed hawks dot the sky.

"They all come here," Becky said, "because we have so much prairie and lots of mice. Bluebirds too—the ranch has been home to a Western Bluebird preservation project to strengthen their numbers."

"And in the oak groves, we have trillium flowers that come up white in early spring. And bright-yellow balsamroot flowers too, which look like sunflowers. They've been decimated by development, which is sad because the threatened checkerspot butterflies rely on them. But the checkerspots are all over my ranch."

She swept an arm through the air as we approached the edge of the oaks, indicating the vast expanse of prairie before us.

"We embrace all life on the ranch—muskrats to elk to butterflies to coyotes. They are able to live in complete happiness with my cattle because I don't let my cows damage the grass by overgrazing. Every good cattle rancher is first a steward of grass."

This type of synergy between cattle and native flora and fauna abound on well-managed farms.

"This is just how ranches work," Becky said. "People just don't realize it."

*   *

The case is similar at Colvin Ranch a few miles away, where Fred and Katherine Colvin are the fifth generation to farm their plot of Washington State's South Sound Prairie.[7]

When we visited the small Black Angus grass-finished farm near the town of Tenino, Fred told us about the exemption to the Endangered Species Act he'd received for his (environmental) management practices.

A few years back, the U.S. Fish and Wildlife Service listed the Mazama pocket gopher as threatened because all but nine percent of its habitat—Northwestern prairielands—had been decimated. Privately owned lands, including Colvin Ranch, were the last stand for the burrowing critter.

When the Fish and Wildlife Service began notifying Thurston County landowners there would soon be a new "critical habitat" listing, agents who spoke with Fred's USDA conservation-easement advisor and who visited the ranch were blown away by how pristine the prairie ecosystem on Colvin Ranch was.

Fred told us, "It's just how we've been doing it for 150 years."

The impetus for Fred and Katherine to maintain their prairie is economic: A healthy ecosystem keeps land productive over multiple generations, ensuring their grandchildren too will be able to make a living farming. It wasn't just a happy coincidence that the outcome was positive from an environmental perspective. For small-scale ranchers, environmental sustainability and economic viability are inextricably intertwined.

And the quality of the beef isn't unrelated to the equation either. In our own travels, we've found that the most well-managed farms correlated with the most fantastic-tasting steaks. To this day, Colvin Ranch's grass-finished New York strip steak is one of the most tender pieces of beef we have ever eaten, period. It was like cutting into a down pillow, and as mild and gentle in flavor as the best grain-finished steaks but with an attractive hit of citrus to accompany the light malty flavor.

* *

Back at Thistle Creek Farms with the Lakes, Christy joined me where I stood leaning against the gate.

She gestured out over the thigh-high grasses. Some of them, topped with little, many-petaled balls, I could identify as clovers, but the rest were a mystery.

"We planted grasses that are very valuable," she said, explaining that over the years, George has imported his grass seeds from other countries, including Denmark, Romania, New Zealand, Ireland, and the Czech Republic, though now a company called King's AgriSeeds grows some of his preferred grasses on U.S. soil. All in all, George's grass strategy is considerably more complex than most.

"It took an investment," Christy said.

"So what we're doing here with paddocks," she continued, "is part of not ruining that investment. If you buy good grass, you better move your cows."

What she meant by that, George told me later by email, is that it's important not to overgraze.

"We want the plants to have time to heal and begin growing," he wrote.

As I looked out toward the land into which Christy and George had poured so much money, I thought about something Christy had mentioned earlier in an offhand sort of way as we sipped our lattes: She and George were giving up on a huge payout by choosing to keep their land.

One neighboring farmer had just sold his ranch to a developer planning to subdivide the property. State College commuters, apparently, were clamoring for more housing, and the verdant rolling hills of Huntingdon County were an appealing prospect—if the ranchers occupying them would take the cash.

But the Lakes have chosen to keep Thistle Creek Farms in the family.

"It's just the right thing to do," George said.

The emotional attachment to their land is what gives ranchers like them a more compelling reason to conserve land than most other private landowners in the country.

* *

George reappeared, walking back up from the stream, two plastic stakes in his hand and a growing roll of white plastic cord—the temporary fence—wrapped around his forearm. He was opening up access to a new paddock incrementally, but I still couldn't see any cows.

Leaning against the fence and shading her eyes with a hand as she too squinted fruitlessly, Christy told me the herd was lazing in the shade below the edge of the forest a few acres away.

She called to George, who had a better view than us, halfway down the field, "Are they coming?"

THE FETCH[8]

A poem by George Lake

The Van Gogh colors of fall are gone.

Replaced by the grey tweed of winter.

But, this early morning blanketed by stark, blue white of
deep powder.

I shiver on my horse, hands too cold to hold the reins.

A half mile off, three hundred cattle lie like black granite
on the distant ridge.

I whistle them out, through trackless snow.

Leaping like bulls from a bucking chute, they lunge toward
the herd.

I send Hemp left, Taff right.

Each turn causes an explosion of snow like tiny grenades.

The herd, now stretched over a quarter mile seems too
much for two dogs.

Front to back, left to right they work.

Too much passion can burst the heart.

No sound but the muffled beat of many hooves.

The herd approaches, steam exploding from nostrils, forced
on by the fetch.

Cows now settled in new feed.

Dogs biting at ice balls between toes.

> Tongues hanging, defying the cold. For more than a mile
>   their work is written—in snow.
> The arc, the turn, the straight track speak of their passion,
>   their love of the fetch.

"Oh yeah," he replied, grinning. "I don't even need a dog for this."

I glanced toward the forest edge, wondering why the Border Collies were getting a day off. Christy saw the cows before I did.

She said, laughing, "Caroline, look at them coming."

The sound of bells reached my ears, and as I stood on tiptoe in my barely worn-in cowboy boots, I could just make out a group of a few hundred black and red dots appearing over the rise between us and the forest. The cicada-filled morning air was joined by jingling, echoing music as the herd approached the unrestricted, new patch of grass.

"It's like they have a chef in their kitchen," Christy chuckled. "'If George is around, go toward him for food.' They just know. It's easy as calling kids to the table."

The cows bent their heads over the grass as one, the bells at their necks still jingling lightly.

George sidled back up to us then, wiping his forehead. He smiled, clearly having a blast moving the cows. He would do this several more times today, practicing a method that intentionally encouraged microbes and regenerated his soil, yet he looked less like a scientist than a guy giddy after hanging out with his pets.

"You know," he said, coming to stand next to me at the fence, "one time, somewhere at a talk I gave, some animal-rights woman said to me, 'How can you kill these animals?'"

He took off his hat and wiped his forehead.

"I told her," he said, jerking a thumb back over his shoulder at the herd behind us, "if I only had one pet cow, there are 399 other cows in there that wouldn't get a chance to live. And you see how they were bouncing across that hill—they're living a good life."

CHAPTER 6

A FLIGHT OF BEEF

—

NO TWO STEAKS ARE THE SAME

—

Raw mushroom. Something burnt, a bit like red pepper sauce. Wood. Blue cheese. Clean, clear juiciness. The flavor not very buttery, yet tender and utterly mouthwatering all the same. No, I wasn't describing wine, coffee, or chocolate. This was a steak tasting, unexpectedly complete with all the variance of coffee or microbrew, each of the three steaks as distinct from one another as a Zin, a Cab, and a Chardonnay.

You may have taken part in a wine tasting, but you've probably never heard of anyone doing a steak tasting. What you'll find, when you sit down with craft beef of the same cut from a range of different farms, is that it's easier to tell steaks apart than to distinguish between glasses of wine. (And you don't have to spit out the steak between tastes.)

THE ROAD TO CONNOISSEURSHIP
IS PAVED WITH STEAKS

A few months after tipping their first cow, Joe and Ethan decided to throw their very first "steak tasting."

"At that point," Ethan explained when he first told me this story, "we really didn't know all that much. But we thought, 'Let's do a beef version of a wine tasting and see if people can taste the differences.'"

So they invited a mix of Seattle chefs, food writers, and friends over to Joe's house on Mercer Island and prepared to taste beef from three farms: Sweet Grass Farm, Step By Step Farm, and Harlow Cattle Company.

"There was relatively little method at that point," Ethan said. "We weren't even trying to hold the cut of beef constant across the farms."

"And of course the chef we'd arranged to have cook the beef," he added, laughing, "his car broke down on the side of the road, so I ended up manning the grill, and I burned the tri-tip."

The flagrated tri-tip replaced with other cuts, Ethan and Joe served their guests and handed out cards so they could write down words to describe what they tasted.

Joe and Ethan at Magnolia Cattle Company in Bothell, Washington.

"We didn't even prompt them with much really," Ethan said. "But they all kept coming to the same word when they tried the Wagyu slider. It universally got called 'buttery,' much more so than the others," he went on.

"I think it was a moment of surprise for a lot of people. Because if what you're used to is commodity beef, well every element of that system—from the 90-day steamed-grain feedlot ration to the mechanical tenderization process—is designed to wipe out variances in flavor and ensure absolute uniformity. So everyone was sort of just standing there, looks of wonder dawning on their faces, as they realized that beef—craft beef anyway—doesn't all taste the same."

The experience boosted Ethan's and Joe's confidence that they were on to something.

Their next foray into steak tastings had slightly more method behind it. A few months later, local chef Tom Douglas lent them a proper kitchen, they invited more Seattle chefs, and made sure to represent six farms' worth of beef.

"We knew a little more at that point, so were going for breadth of flavor," Joe said. "We didn't do a very good job of keeping it apples to apples though."

Even so, they came out of the experience with a slightly more developed vocabulary.

"That time, people came up with words like *meaty*, *tangy*, *earthy*, *beefy*. And those words," Ethan explained, "became important reference points for us."

So when, in the summer of 2016, George Neffner at Magnolia Cattle Company outside Seattle called them up to propose a Wagyu tasting, they were ready.

George was raising fullblood Wagyu cows, but instead of

keeping them in roomy pens in the traditional Japanese style, he was raising his Wagyu herd on grass pasture and finishing them on grain.

A5 Wagyu cooks on a Himalayan salt block.

George suggested they compare several Wagyu varieties: A5 Wagyu imported directly from Japan, George's own fullblood Wagyu raised on Seattle soil, the fullblood Wagyu of two other American ranchers, and the grass-finished purebred Wagyu that Scott Meyers was raising on Lopez Island.

"At that point," Joe said, "we understood that you have to keep the cut constant if you're trying to see how another variable, like the farm it's from, affects the taste, because flavor and texture vary a lot by the cut.

"Cut is a whole additional dimension that wine doesn't have—you never compare the fruit and the skins and the stems of a grape, do you? But if you did, that would add much more complexity to the concept of a wine tasting. And that's what you have with beef, on a grander scale: A cow, after all, is an infinitely more complex organism than a grape."

So Joe, Ethan, George, and some friends all gathered around George's Himalayan salt block—a square, pink slab of mined salt that, when superheated, makes for a dramatic surface for cooking steak—and wondered the same thing:

If you really did this right—which is to say, held as many "variables" in the beef equation constant as you could—would you be able to distinguish the flavor of various Wagyus by the feed it had received? Or by the activity level of the cows?

Joe was surprised by what he found.

"They did actually all taste different," he said. "There were subtle differences. Like, I could feel that I was new to this . . . and someone more experienced could say exactly what the differences actually *were*.

"But even for me as a beginner," he added, "it was apparent that one steak was immediately beefier than the other. Another had the edge on sweetness. Another left a different taste in your mouth, and so for the first time, I started to appreciate the aftertaste as an element of the flavor experience."

The least "beefy" of the collection, he reminisced, was the Japanese A5 Wagyu.

"It more than all the others had a pure, sweet tone, almost like pâté, while the fullblood Wagyu raised on pasture and grain finished," Joe said, "were noticeably beefier."

Ethan chimed in, adding, "And that makes sense, because they would have spent time on pasture, developing muscles. You expect more of the flavor that we call 'beefiness' when animals are using their muscles a lot."

His favorites, to his surprise, were the Magnolia and the Sweet Grass, with bolder, lingering flavor.

"I loved them because they were unapologetically beefy."

* *

Inspired by Joe and Ethan, I decided to hold my own steak tasting. But I wanted to do a little prep work to hone my chops first, so I consulted two experts.

One was my college friend Robert, who told me over the

phone how the practice of "cupping" coffee, which he learned during his stint working at Nashville's Barista Parlor, informed his quest for the perfect steak at his family's own Angus beef farm outside Memphis.[1]

On free evenings, Robert had taken to cooking two steaks, each from a different one of his family's grass-fed, grain-finished Angus-Hereford steers, and tasting them next to one another to identify differences.

"Beef isn't a homogenous product," Robert explained, "so you get a lot of variability between steaks from different areas, or from different farms. Even within a farm, the flavor can be hard to control, because you'll have fields with more clover in them, or fields with weird grasses, and some cows honestly have preferences."

"For example," he said, "Sally might like clover, and Suzie might like fescue. So depending on which you're eating, your steak might taste different from one day to the next."

Robert's goal was to find those flavor differences and troubleshoot them backward to the pasture level to create a fairly consistent product at the butcher counter. That's where his coffee training came in.

"At Barista Parlor," he explained, "I would taste just two at a time, where only one variable is different between them. That's important when you're starting."

Eventually, Robert's palate for coffee was so refined that he could take a sip from two different cups, make a note of flavors like Meyer lemon and brown sugar, and then flip the bag over and find—lo and behold—"literally those words listed as tasting notes on the back. That was unreal."

He was still working by night toward that level of palate development with beef, so I wasn't too far behind. His point was helpful: Start simple. Compare just two steaks, and see if you can identify their differences.

That brought me to my second expert, Kurt Beecher Dammeier, founder of Beecher's Handmade Cheese and Seattle's resident expert on the craft-food revolution.

*   *

One sunny Seattle day, Kurt and I met at his airy steakhouse The Butcher's Table in South Lake Union.[2]

Someone had had the good sense to throw open the windows, and music spilled out onto the sidewalk near the table where we sat talking—and tasting—steak.

Kurt told me that he'd had a lifelong passion for beef and had fallen head over heels in love with Wagyu-Angus cross a few years prior when it was featured as a special at one of his own restaurants.

He couldn't shake the feeling that beef might be heading for a craft explosion not unlike the transformations he'd watched happen in his fields of specialization, cheese and beer. And so he had decided to open a steakhouse focused entirely on Wagyu-Angus cross.

But the focus on craft beef wasn't the only thing that made The Butcher's Table special. Having bought all the animals that would ever be converted into steak for his restaurant, Kurt was bringing to life a vision for a totally new kind of steakhouse: one that knows about the cows from which the steaks come and can vouch for their humane treatment.

We compared Kurt's Mishima Ultra[3]—the most marbled of the cross-beef he raised, which would rate well above Prime—to a grass-finished Angus-Hereford steak from one of my favorite farms in Washington State, Colvin Ranch.

Unsurprisingly, seeing as he was a Wagyu devotee, Kurt preferred the Wagyu-cross, but he liked the New York strip steak from Colvin too.

The steaks Kurt and I tasted: the before shot and the just-before-we-devoured-them shot.

"It's a lot more tender than I thought it would be," he acknowledged approvingly as he cut off another strip and took a bite.

I loved both—the slight umami flavor of the Wagyu-Angus was delectably savory—but slightly preferred the Colvin steak and its citrusy, almost oystery aspect. I was biased. Of all the steaks I'd tried in the past year, Colvin had become my true favorite.

The truth was that both steaks were phenomenal but completely different.

I was ready, now, to throw a true tasting.

THE (SEMI) BLIND TASTE TEST

One evening after work, I decided to taste test New York strip steaks from three different Pacific Northwest farms. Represented in the lineup were Novy Ranches, a grass-finished Angus beef farm in the shadow of California's Mount Shasta; Hutterian Farm, a pasture-raised, grain-finished Angus beef farm in Washington State specializing in no-till farming and homegrown corn and peas; and Sweet Grass Farm, the grass-finished purebred Wagyu farm on Lopez Island that I'd visited earlier in the year.

Having finally become a marginally competent preparer of steak, I decided to cook the flat irons—each about an

FOUR OF THE FACTORS THAT
GO INTO STEAK FLAVOR

Feed matters. And not just grass versus grain, but what *types* of grasses and what *types* of grains (and legumes and fruits, for that matter!) along with the time of the season at which the plants were harvested. Any grass-finished beef farmer worth his or her salt speaks in "haylage," "baleage," and "forage curves," and will tell you they couldn't just mow the lawn willy-nilly and feed cows the scraps to produce good beef—that's exactly how you get the "bad," super-tough grass-finished beef that tends to give *all* grass-finished beef a bad rap.

On the other side of the feed coin, a diet of steamed and flaked corn on a concentrated feedlot produces predictable but unexciting flavor compared to the farms that raise homegrown, chemical-free grains like millet or who add apple pomace from their own orchard to their herd's feed. Industrial beef, ultimately, is interested in squeezing out costs, which means they opt for the highest-volume, highest-energy grain available to pack on fat: corn, soy, and other processed grains like distillers grains, a byproduct of ethanol production. Because feedlots are paid by the pound, and because marbled beef is in such high demand, Prime beef catches the highest premium. In the industrial system, quality of feed, animal health, and even flavor take a backseat.

Breed matters. Wagyu beef from the Kuroge Washu breed represents a noticeably more marbling-prone gene pool than, say, Angus beef. But it gets even more complicated than that, with some breeds having a better predisposition to marble on a grass diet than other breeds, like the French breeds Charolais, Maine-Anjou, and Limousin. In America, the Angus-Hereford cross is particularly popular among small-scale ranchers for combining the meat quality of Angus with the mellow temperament of Hereford. In very real terms, temperament matters to ranchers because with chill animals, you're less likely to get trampled by your herd, and their meat is likelier to be tender, because they're stress-free.

Age matters. Cattle harvested younger tend to produce more tender meat and have less-developed flavor. European cattle producers let their cows grow longer to let flavor deepen and become more complex. The USDA has a slaughter cut-off of 30 months due to lingering fears over bovine spongiform encephalopathy (BSE), so if cattle are slaughtered later than 30 months, the spinal cord and brain tissue have to be removed entirely in order for the rest of the cow to be used. To get around the hassle, most ranchers and feedlot operators in the U.S. choose to grow their cattle a shorter time period—some pushing right up to the twenty-ninth month without going over. CAFO beef can be as young as 12 months, which some eaters consider to be more like veal because of its mildness and lack of flavor.

Soil matters. Soil is filled with all sorts of nutrients that vary according to what's present in a given region, like limestone formations that give rise to bioavailable calcium, which in turn help cows grow strong and healthy. On top of that, various nutrients and vitamins may or may not be available for uptake by plants depending on the pH of the soil. Legumes like alfalfa require a pH of 6.5 or slightly higher to grow, whereas some grasses can survive at a pH as low as 6.0. Northeastern soils tend to be more acidic, for example, and thus require the application of a "soil amendment" like ground limestone in order to be able to grow grasses and legumes.

Handling matters. A stressed-out cow is known to produce tough beef. That's exactly how you get "dark cutters." The beef industry's nightmare outcome, dark cutters (also known as DFDs for "dark, firm, and dry")—whose meat is literally a dark purplish-red-brown color due to excess stress before slaughter—are a near-total loss because consumers have historically rejected dark meat when shopping at retail stores. The USDA doesn't ban dark cutters from entering the food supply, but that beef never makes it to retail stores, and they don't earn a packer the full price of Select or Choice meat.[4] That's why farmers who know what they're doing emphasize a "zero-stress" management approach—the happier the cow the more tender and delicious the steak.[5]

inch thick—in the oven at low heat to move their internal temperature a hair past rare and then sear them quickly on my cast-iron skillet to give the ever-delicious Maillard reaction time to stretch its legs.

WHAT IS THE MAILLARD REACTION?

The Maillard reaction is that beautiful occurrence when your steak contacts extreme heat—like a heated cast-iron pan or a direct flame—and a delectable crust is produced as the meat sears. Scientifically speaking, the sugars and proteins on the surface of the steak are being transformed, producing new flavors: beefy, malty, charred. There is no salivating cavewoman without the Maillard reaction. "Boiled" flavors just aren't that tasty.[6]

Having thawed in the fridge all day, the steaks were cold but ready to be removed from their plastic wrapping. I noticed they were different in coloration, and realized my attempt to keep the tasting blind was probably going to be futile.

The Hutterian steak was richly marbled—it would have graded Prime easily—and its color was a lovely, bright crimson. The Novy steak—grass finished, I remembered—was a darker red, the fat not white but a light beige. The Sweet Grass steak was show-stoppingly dark-hued. It hovered somewhere between blood red and rich, chocolate brown. It was also visibly juicy, yet this particular New York steak of Sweet Grass's was not very visibly marbled save a few dark-cream flecks, which surprised me. But then again, I remembered, thinking back to my conversation with Robert Schutt, even at the farm level, steaks can differ drastically.

I used my coworker Jacob's favorite method to cook them, preheating the oven to 150 just to warm them throughout. That would be followed by a quick, 450-plus-degree sear on a cast-iron skillet.

They warmed quickly—my digital thermometer registered 115 degrees in each steak within seven minutes.

As soon as the surface of the cast iron hit 450, I placed each steak on the pan and cooked 90 seconds each side. Now they were hovering around medium rare—known to be the optimum level for tenderness for the flat iron.

As I was dutifully letting the steaks rest, I heard a key turn in the lock and my housemate, Mandy, walked in.

Mandy and I had attended college together. She'd migrated to Seattle later than me, after finishing a Mexico-to-Canada hike of the Pacific Crest Trail. Like me, Mandy had spent much of her life as a vegetarian. Unlike me, she had not yet renounced her ways.

She had, however, been hearing me spout animatedly about beef for months. I suppose our conversations had awakened a curiosity or steak nostalgia in her, because when she walked into the kitchen and saw the three steaks resting on plates distinguished by Post-its marked A, B, and C, she asked, "Would you mind if I tried them?"

The chef and beef nerd in me beamed. Was I, the former vegetarian, about to convert *my* first vegetarian?

Mandy tasting alongside me also solved my *it's-not-blind-anymore* problem—the taste evaluation would be completely blind for her, even if it wasn't for me.

I pulled out two plates from our cabinet, sliced strips of each steak, and put three on each plate. I took my first bite of the Hutterian steak as Mandy watched, looking excited but slightly apprehensive.

I chewed, focusing every ounce of my mental effort on exactly what I was tasting. I was trying to improve my palate here, and Mandy's presence made me feel a little pressure.

On the tenderness scale, I noted, it was fairly tender but on the lively side of things—definitely not mushy. Maybe closer to bouncy? I felt a little silly flexing my steak-vocabulary muscles, but *tender* just didn't quite do the experience justice.

The Hutterian steak was juicier than you would think a piece of meat could possibly be. It positively oozed juice, yet it wasn't the slightest bit oily, like most grain-finished beef is. It was clean and calm, leaving—like I aspire to when I go hiking— no trace. (Ha.)

Trying to channel the sommeliers I'd watched swirling glasses in Netflix documentaries, I next turned my attention to the character of the steak and alighted on three words: simple, straightforward, unfussy. The flavors, finally, were parmesan and something distinctly, roundly sweet, like melon.

I said some of this last bit aloud to Mandy, and she looked at me sideways.

"You can really tell all that?" she asked. "*Melon?*"

"Well . . . to a certain extent," I replied, which was true. I did, of course, feel like a bit of a cigar-swilling, tweed-wearing foodie saying "this steak has notes of melon." But it actually did.

I told Mandy how, a year ago, I would never have even considered that you could pick out distinct flavors in a single piece of steak. Steak, I thought back then, was steak. (Was steak, was steak, was steak.) But no longer! I knew steak now as something to be understood, enjoyed, divined, considered, adventured through.

Making the smallest fist-clenching motion, as if steeling herself, Mandy took a step toward the counter and picked up the Hutterian steak (or, to her, the *A* steak).

She chewed slowly. I knew a brief moment of panic when I wondered if I'd led this vegetarian straight down into a moral hell.

But then, to my relief, she grinned.

"That's really good."

I laughed. "Isn't it? Wait until you try the others."

I speared the *B* steak—really Novy—on a fork.

This was different, I could tell, but I didn't know quite how. I paid more attention. It was . . . slightly grassier? No, that was wrong. What I was tasting wasn't grass; it was spinach. And walnuts. And then beefy flavor. But between that, little hints of that spinach again. It was firm but not chewy. It, more so than the grain-finished steak, I noticed with surprise, coated my tongue. Not unpleasantly so. The flavors came and went like waves.

Mandy looked like she was trying to puzzle a difficult math equation.

"Well, it's definitely not the same."

I prompted her. "What do you taste? Say random words that come to mind."

"Uh . . . " she blanked, then said, "It tastes somehow less . . . well less what I expected. It's a little harsher, but I don't mean that in a bad way. I don't know. It's really great. It's different."

I raised my eyebrows. She was good. First of all, she wasn't freaking out over her first bite of meat in a decade. She was going through the motions of a tasting. What a star. She was also doing way better at divining the differences than I had at the beginning.

The final strip to sample was the Sweet Grass, which was the Wagyu breed but raised on American grass pasture. Its maroon hue, even cooked, was still as dark and deep as cranberry sauce. I popped it into my mouth, and—

"*Wow,*" I breathed.

Mandy mimicked me, and her eyes grew wide. "Jeez," she commented.

I knew exactly what she meant. These flavors jumped out at me without preamble. Raw mushroom. Something burnt, a bit like red pepper sauce. Wood. Blue cheese. Juiciness like nothing I'd ever tasted. Not entirely buttery, but tender and utterly mouthwatering. Still clean and oilless, somehow. This steak, unlike either of the other two, tasted like it had plodded straight out of the pasture.

Mandy had slid down against the dishwasher, still chewing.

"Cut more," she instructed.

* *

As Joe likes to say, if you're having a steak dinner party, it's quite likely you've paid more for the steak than the wine. So why is it, then, that you know exactly where the wine came from—probably even a little something about its varietal, its maker, and its vintage—yet you know little to nothing about the steak? Steak tastings celebrate what's different about farms and flavors across America. And even more importantly, they let you figure out what kind of beef you like.

We recommend that for a true blind tasting, you should start with beef that is all of the same cut. Traits like marbling, tenderness, and nutrient content vary greatly from cut to cut. Your steak tastings need not all revolve around New York strips or ribeyes—that could get a little pricey, and actually might be boring too. Why don't you try comparing oxtail from several different farms, each braised for the same length of time? Ground beef is considered to be a good control cut with which to hone in on flavors, but so is tongue—same texture if you chop it up, but much more adventurous.

The sweet spot is to taste beef from about three to five different farms. Many more than that and your taste buds will be working in overdrive. When he's hosting a more formal tasting for friends, Ethan will have everyone eat a slice of green apple between steaks to clear their palate. Three farms will give you a spectrum of flavors.

If you're approaching this steak tasting like a true beef nerd, hold all variables constant but one. Yes, you could call this a fool's errand. There are a million potential variables in the craft-beef equation, and thus a million small ways you can end up with different-flavored steaks, but you *can* keep *some* variables constant. Say you're interested in determining whether there are regional differences in the flavor of beef. Track down three farms in, for example, Florida, Washington State, and Pennsylvania, and make sure they're all the same 1) breed, 2) age, and 3) feed. If you're really willing to do your homework, call up the farmers and ask about the types of grasses and grains, to keep the feed *really* constant. A corn-based feed, for example, tends to taste a little different than a millet-based feed. All grass and all grain are not created equal.

A few final tips: Steak tastings are more fun with company. Invite friends. And don't forget to label your steaks. It really sucks if you mix them up. Eat and be joyful!

CROWD COW

How to host a steak tasting

Put all your new beef knowledge to use and host a steak tasting! To get a sense for variety of flavors, compare beef from 3 farms, holding the cut constant. For a blind tasting, nominate a Steakholder to know which steak matches which farm and to serve guests steaks A, B, and C.

Steak Profile - fill this section in only after tasting is complete

Farm ....

Cut: .... **Feed + Finish:** ....

Breed: .... **Butchery:** ....

Tasting Notes - use the word clouds to guide your review of the taste

TEXTURE

Firm, Soft, Chewy, Tender, Very firm, Velvety, Silky, Squeaky, Fibrous, Tough, Sinewy, Rubbery, Gristly

....

....

....

MOUTHFEEL

Clean, Buttery, Oily, Greasy, Powdery, Juicy, Dry, Rich, Round, Party in your mouth

....

....

....

FLAVOR

Creamy, Grassy, Mineral, Prune, Malty, Beefy, Gamey, Smoky, Citrusy, Plum, Char, Steely, Nori, Butter, Oysters

....

....

....

INTENSITY

○——○——○——○——○

gentle   very strong

OVERALL OPINION

○——○——○——○——○

'meh   steak of my dreams

For a printable version of this tasting chart, go to www.crowdcow.com/tasting_chart/

ACKNOWLEDGMENTS

___

This book would not have been possible without the help, guidance, and inspiration of many brilliant people. First, to the hardworking family farmers mentioned in and interviewed for this book—George and Christy Lake, Fred and Katherine Colvin, Bob and Kate Boyce, Paul and Kristin Uhlenkott, among many others—thank you. To Scott Meyers at Sweet Grass Farm and Becky Harlow Weed at Harlow Cattle Company, in particular, we appreciate you taking us under your wings and sharing many conversations about forages, small-farm economics, and what it takes to produce a great steak. George Neffner, thanks for having us out to Magnolia Farms for our first-ever use of a Himalayan salt block and foray into the intricacies of Wagyu.

To Jay Young, Brad McCarley, Tim at B&E Meats in Queen Anne, and Jim Carlson and David Barney at Minder Meats in Bremerton, thanks for telling us which end of the cow is which and what little-known cuts are great to eat. Chefs Brock Johnson, Taylor Thornhill, Kurt Dammeier, Billy Oliva, and Thierry Rautureau also guided us toward especially delicious cuts and meals. Thierry, that foie-gras burger at Loulay was wild. Eric Hellner, you really know how to talk—and serve—Japanese Wagyu. Mark Canlis, thanks for the late-night Jungle Birds and tiki cups. John Howie, we never could have seared A5 perfectly without your guidance. And Tom Douglas, thanks for lending us your space and making our first-ever steak tasting possible, in the earliest days of Crowd Cow.

Eric Koester and Brian Bies proved invaluable editors and guides throughout the process of writing our first book. The edits and feedback of Robert Lowry, Elizabeth Liu, Jacob Tally, and Mandy also pushed us toward a better manuscript, and for that we'll always be grateful. Jay Herratti and Jeremy Price helped us cross the finish line. Andrea Burnett was a fierce supporter, advocate, and representative for all things *Craft Beef.* Joe Montana believed in us while most people thought our idea was still a joke. Thank you.

This book is, in some ways, the product of years and years of contemplating and drooling over food, especially steak. The writers and thinkers who inspired us were many: Michael

Pollan, Mark Bittman, Marion Nestle, M. F. K. Fisher, Tom Steyer, Bill Cronon, Dan Barber, Barbara Kingsolver, Mark Schatzker, and Allan Savory, to name a few. Kenji Lopez-Alt and Harold McGee introduced us to the science behind the deliciousness of a medium-rare steak. Steven Raichlen's writings taught us how to dry-age beef. Michelle Tam helped us hone our philosophy on what it is to eat wholesome food, while Sarah Ballantyne reminded us the importance of knowing where your food comes from. Anthony Bourdain made us realize the sin of failing to let a steak rest (thank you, Anthony, we will remember). Francis Milman, David Chang, Alice Waters, Wolfgang Puck, Tom Colicchio, Jamie Oliver, Thomas Keller, and Gordon Ramsey are our culinary heroes. Meathead Goldwyn, Matt Pittman, Clint Cantwell, Aaron Franklin, and Jess Pryles are our barbecue heroes. Big Green Craig, Myron Mixon, and Grillin' Fools were vital parts of our (ongoing) barbecue and grilling education too.

Ree Drummond, Kevin Curry, Sebastian Noel, Lisa Leake, Gina Homolka, Daniel Vaughn, Juli Bauer, your recipes and writings inspire us. Beth Kirby, your photos nearly make us pass out, they're so beautiful. Ashton Kutcher, we're glad you're bringing American viewers to the ranch. It's about time we trained the camera there.

Nick Offerman/Ron Swanson, you had us at meat tornado. Then you stole our hearts at "turf and turf." To Carrie

Brownstein, thank you for the scene with Colin the chicken. It will live in infamy. And finally, to Brendan, thank you, thank you, thank you for walking into Joe's office and saying the words that would launch a new chapter in our lives: "I'm getting my cow on Friday!"

APPENDIX I

**RESTAURANTS IN AMERICA OFFERING
AUTHENTIC KOBE BEEF, AS OF FEBRUARY 2018:**

There are 25 restaurants in America that sell authentic Kobe beef.[1] They don't necessarily all have Kobe at all times, so be sure to check the menu listings or give the restaurant a call before heading there expressly for Kobe!

---

[1] "Shop List." 2018. *Kobe Beef.* Accessed February 1. http://www.kobe-niku.jp/shop/?lang=1&prefecture=&tag=3. This is the official list of restaurants licensed to sell authentic Kobe in the U.S.

CALIFORNIA

- Alexander's Steakhouse, 19379 Stevens Creek Blvd, Cupertino, CA 95014
- Alexander's Steakhouse, 448 Brannan St, San Francisco, CA 94107
- Shibumi, 815 S Hill St, Los Angeles, CA 90014
- Arsenal, 715 Brannan St, San Francisco, CA 94103
- Jean-Georges Beverly Hills, 9850 Wilshire Blvd, Beverly Hills, CA 90210
- Omakase, 665 Townsend St, San Francisco, CA 94107
- Roka Akor San Francisco, 801 Montgomery St, San Francisco, CA 94133

NEW YORK

- 212 Steakhouse, 316 E 53rd St, New York, NY 10022
- Castle Hotel and Spa, USA, 400 Benedict Ave, Tarrytown, NY 10591

NORTH CAROLINA

- Meat & Fish Company, 919 S McDowell St, Charlotte, NC 28204

HAWAII

- 鉄板焼 銀座おのでら (or Teppanyaki Ginza Onodera), 1726 South King Street, Honolulu, HI 96826
- The Pacific Club, 451 Queen Emma Street, Honolulu, HI 96813

ILLINOIS

- RPM Steak, 66 W Kinzie St, Chicago, IL 60654
- Gibsons Italia, 233 N Canal St, Chicago, IL 60606
- Roka Akor Chicago, 456 N Clark St, Chicago, IL 60654
- Benny's Chop House, 444N Wabash Ave, Chicago, IL 60611

ARIZONA

- Roka Akor Scottsdale, 7299 North Scottsdale Rd, Scottsdale, AZ 85253

VIRGINIA

- Zoes Steak and Seafood, 713 19th Street, Virginia Beach, VA 23451

TEXAS

- Nick & Sam's, 3008 Maple Ave, Dallas, TX 75201
- B&B Butchers and Restaurant, 1814 Washington Ave, Houston, TX 77007
- B&B Butchers and Restaurant, 5212 Marathon Ave, Fort Worth, TX 76109
- Roka Akor Houston, 2929 Weslayan St #100, Houston, TX 77027

NEVADA

- Bazaar Meat by José Andrés, 2535 S Las Vegas Blvd, Las Vegas, NV 89109
- MGM Resorts International, 3730 Lubio S LU, NV 89149

(including Jean-Georges Steakhouse at ARIA, 3730 S Las Vegas Blvd, Las Vegas, NV 89109)

- Wynn Las Vegas, 3131 Las Vegas Blvd, South Las Vegas, NV 89109

APPENDIX II

All photos herein were taken by and are the property of Joe Heitzeberg or Caroline Saunders, with two exceptions: The good folks at Steaklocker provided us with the image of the dry-aging fridge in Chapter 2, and Becky Weed lent us two wonderful pictures of her and her dog, Tipper.

NOTES

INTRODUCTION

1. Hribar, Carrie. *Understanding Concentrated Animal Feeding Operations and Their Impact on Communities.* Report. Edited by Mark Schultz. 2010. Accessed November 13, 2017. Understanding Concentrated Animal Feeding Operations and Their Impact on Communities. 2010. Accessed November 12, 2017. https://www.cdc.gov/nceh/ehs/docs/understanding_cafos_nalboh.pdf.

   On pages 3–7, the report by NALBOH discusses the impacts of CAFOs on groundwater, surface water, air quality, greenhouse gas and climate change, and odors.

   Milman, Oliver. 2017. "Meat Industry Blamed in Largest-Ever 'Dead Zone' in Gulf of Mexico." *The Guardian*, August 1, 2017. https://www.theguardian.com/environment/2017/

aug/01/meat-industry-dead-zone-gulf-of-mexico-environment-pollution.

2. Guglielmi, Giorgia. 2017. "Are antibiotics turning livestock into superbug factories?" *Science*, September 28, 2017. http://www.sciencemag.org/news/2017/09/are-antibiotics-turning-livestock-superbug-factories.

3. Hribar, *Understanding Concentrated Animal Feeding Operations*, 2010, 3–7.

4. United States Department of Agriculture. 2013. "Feedlot 2011 Part IV: Health and Health Management on U.S. Feedlots with a Capacity of 1,000 or More Head," September, 2013. https://www.aphis.usda.gov/animal_health/nahms/feedlot/downloads/feedlot2011/Feed11_dr_PartIV.pdf. See p 28, Figure C.1.b.

5. Shanker, Deena. 2017. "Trump Chooses Big Meat Over Little Farmers." *Bloomberg*, October 25, 2017. https://www.bloomberg.com/news/articles/2017-10-25/the-trump-administration-chooses-big-meat-withdraws-gipsa-rule.

6. Urry, Amelia. 2015. "Our crazy farm subsidies, explained." *Grist*, April 20, 2015. http://grist.org/food/our-crazy-farm-subsidies-explained/.

7. Lardy, Greg. Feeding Corn to Beef Cattle. Publication no. AS1238. Animal Sciences, North Dakota State University. Accessed October 1, 2017. https://www.ag.ndsu.edu/publications/livestock/feeding-corn-to-beef-cattle.

Corn is the preferred grain for finishing grain-finished cattle in the United States, both because of its low cost and because of its high energy content. Corn is about 70 percent starch on a dry-matter basis.

8. 8. "Navigating Pathways to Success." 2016. *National Beef Quality Audit*. Beef Checkoff. https://www.bqa.org/Media/ BQA/Docs/2016nbqa_4-pager.pdf.

CHAPTER 1: HAPPY COWS TASTE BETTER

1. "Our fleet." 2018. *Washington State Department of Transportation*. Accessed January 30. https://www.wsdot. wa.gov/Ferries/yourwsf/ourfleet/.

    The name Yakima means, in the Yakama language, "To become peopled; black bears; runaway; people of the narrow river." Tillikum comes from the Chinook dialect, and means "friends, relatives."

2. Pollan, Michael. 2006. *The Omnivore's Dilemma*. New York, NY: The Penguin Press.

    In this passage on page 307, Pollan is reflecting on an essay by the English writer John Berger titled "Why Look at Animals?"

3. Meyers, Scott, and Brigit Waring. "Interview at Sweet Grass Farm." Interview by author. August 16, 2017.

4. "Wagyu around the World—USA." 2017. *Wagyu International*. Accessed August 12. http://www.wagyuinternational.com/ global_USA.php.

    This webpage is a great resource on the history of Wagyu exports from Japan. The estimate of 36,000 fullbood and purebred Wagyu cattle in the U.S. is based on Wagyu

International's 2010 assessment that there are 40,000 Wagyu (including crosses) in the country. A bar graph depicts that of those 40,000, approximately 90 percent are either full-blood or purebred.

5. Schatzker, Mark. 2010. *Steak: One Man's Search for the World's Tastiest Piece of Beef*. New York, NY: The Penguin Group.

This is the book that made Caroline, a vegetarian at the time she read it, fall in love with meat.

6. "Mobile slaughter/Processing units currently in operation." 2016. *eXtension*. March 20. http://articles.extension. org/pages/19781/mobile-slaughterprocessing-units-currently-in-operation. "Overview of the United States Cattle Industry." 2016. *USDA*. National Agricultural Statistics Service (NASS), Agricultural Statistics Board, United States Department of Agriculture (USDA). June 24. http:// usda.mannlib.cornell.edu/usda/current/USCatSup/ USCatSup-06-24-2016.pdf.

Thistlewaite, Rebecca. "January Email Correspondence." E-mail interview by author. January 30, 2018.

The Island Grown Farmers Cooperative (IGFC) mobile unit was the first USDA-approved mobile slaughter unit in the country. Mobile slaughter represents only an infinitesimal fraction of the total cattle processed annually in the U.S. In 2015, 28.8 million head of cattle were slaughtered; Rebecca Thistlewaite, Program Manager for the Niche Meat Processor Assistance Network, estimates that fewer than 10,000 are killed by mobile slaughter each year.

It's worth mentioning that mobile slaughter isn't the only David against the Goliath of industrial-scale slaughterhouses. Well-run, small-scale slaughterhouses still exist here and there, though many are struggling to remain viable. Those small-scale slaughterhouses, like Rising Spring Meat Company in Spring Mill, Pennsylvania, typically foreground humane treatment of animals and often have long-standing relationships with the same set of small-scale farms.

7. Chambers, Philip G., and Temple Grandin, comps. *Guidelines for Humane Handling, Transport and Slaughter of Livestock.* Edited by Gunter Heinz and Thinnarat Srisuvan. Report no. RAP Publication 2001/4. Regional Office for Asia and the Pacific, Food and Agriculture Organization of the United Nations.

Chapter 2 discusses the effects of stress and injury on meat quality. See: http://www.fao.org/docrep/003/x6909e/x6909e04.htm

8. "National Beef Quality Audit Executive Summary." 2017. Rep. *National Beef Quality Audit Executive Summary.* Beef Checkoff. Accessed October 8. https://www.bqa.org/Media/BQA/Docs/2016nbqa_es.pdf.

The executive summary of the 2016 National Beef Quality Audit report is a helpful source for learning about dark cutters and the relationship between animal stress and meat.

9. Grandin, Temple. "The Effect of Stress on Livestock and Meat

Quality Prior to and During Slaughter." *International Journal for the Study of Animal Problems* 1, no. 5 (1980): 313–37. Accessed December 13, 2017. http://animalstudiesrepository. org/cgi/viewcontent.cgi?article=1019&context=acwp_faafp. Temple Grandin is a preeminent expert on livestock animal welfare in the United States. She's authored many of the studies and articles that we found informative when learning about the relationship between stress and meat quality, and about conditions (and ways to improve) slaughterhouses.

10. Miller, Mark. 2017. "Dark, Firm and Dry Beef." *Beef Research.* The Beef Checkoff. Accessed November 12. www.beefresearch.org/CMDocs/BeefResearch/Dark,%20Firm%20 and%20Dry%20Beef.pdf. A helpful resource for understanding "dark cutters." 1–3.

11. Ibid, 1.

12. Chambers and Grandin, *Guidelines for Humane Handling,* 2.

13. *National Beef Quality Audit Executive Summary,* 14.

14. United States. Department of Agriculture. National Agricultural Statistics Service. *Livestock Slaughter 2016 Summary.* April 2017. Accessed September 4, 2017. http:// usda.mannlib.cornell.edu/usda/current/LiveSlauSu/ LiveSlauSu-04-19-2017.pdf.

   30.6 million head of cattle were commercially slaughtered in 2016.

15. Maday, John. "The Feedlot Death Loss Conundrum." January 25, 2016. Ag Web. Accessed December 20, 2017. https://

www.agweb.com/article/the-feedlot-death-loss-conun-
drum-naa-drovers-cattlenetwork/.

Professional Cattle Consultants (PCC) analyst Shawn
Walters estimated feedyard mortality rates at 2 percent—and
the most common causes, respiratory diseases— when he
was quoted in an AgWeb article.

16. The Center for Genetics, Nutrition and Health. "The importance of the ratio of omega-6/omega-3 essential fatty acids."
Accessed February 1, 2018. https://www.ncbi.nlm.nih.gov/
pubmed/12442909.

17. Ibid

18. Nuernberg, Karin, Gerd Nuernberg, Klaus Ender, Stephanie
Lorenz, Kirstin Winkler, Rainer Wickert, and Hans
Steinhart. "N-3 fatty acids and conjugated linoleic acids
of longissimus muscle in beef cattle." *European Journal
of Lipid Science and Technology* 104, no. 8 (August 19,
2002): 463-71. Accessed February 1, 2018. doi:10.1002/1438-
9312(200208)104:8<463::AID-EJLT463>3.0.CO;2-U.

In this study, German Simmental bulls grazing outside
on grass pasture developed a n-6/n-3 ratio of 1.3, and
those kept in a stable and fed a concentrated diet had a
much higher n-6/n-3 ratio of 13.7, which isn't great for
human health.

CHAPTER 2: SELECT—CHOICE—PRIME—CRAFT

1. Saunders, Caroline, and Billy Oliva. 2017. Interview at
Delmonico's. Personal.

All quotes hereafter are reproduced from Saunders's interview notes and have been confirmed for accuracy by Chef Oliva.

2. Galarza, Daniela. 2016. "A Name You Should Know: Marie-Antoine Carême." *Eater.* June 3. https://www.eater. com/2016/6/3/11847788/careme-chef-biography-history.

3. Reichl, Ruth. 1994. "For Red Meat and a Sense of History." *New York Times*, January 21, 1994. http://www.nytimes. com/1994/01/21/arts/for-red-meat-and-a-sense-of-history. html?pagewanted=all.

4. McGee, Harold. 2011. "The Science of Taste Or: Why Dry-Aged Meat Is So Damned Delicious." *Gizmodo.* December 9. https://gizmodo.com/5866754/the-science-of-taste-or-why-dry-aged-meat-is-so-damned-delicious.

5. Dashdorj, Dashmaa, Touseef Amna, and Inho Hwang. 2015. "Influence of specific taste-Active components on meat flavor as affected by intrinsic and extrinsic factors: an overview." *European Food Research and Technology* 241 (2): 157–71. doi:-doi: 10.1007/s00217-015-2449-3.

6. Nishimura, Toshihide, Mee Ra Rhue, and Hiromichi Kato. 1988. "Components Contributing to the Improvement of Meat Taste during Storage." *Agricultural and Biological Chemistry* 52 (9): 2323–30. doi:https://doi.org/10.1271/bbb1961.52.2323.

7. Dashdorj, Amna, and Hwang, "Influence of taste components," 2015.

8. Lam, Francis. 2013. "Dry-Aged Beef Is a New Trend in

Restaurants Around the Country." *Bon Appétit*, June 25, 2013. https://www.bonappetit.com/test-kitchen/ingredients/article/dry-aged-beef-is-a-new-trend-in-restaurants-around-the-country.

9. Mylan, Tom. 2010. "Dry vs. Wet: A Butcher's Guide to Aging Meat." *The Atlantic*, April 6, 2010. https://www.theatlantic.com/health/archive/2010/04/dry-vs-wet-a-butchers-guide-to-aging-meat/38505/

10. McGee, "The Science of Taste," 2011.

11. Lam, "Dry-Aged Beef Is a New Trend," 2013.

12. Parrish, Jr., F.C., J.A. Boles, R.E. Rust, and D.G. Olson. 1991. "Dry and Wet Aging Effects on Palatability Attributes of Beef Loin and Rib Steaks from Three Quality Grades." *Journal of Food Science* 56 (3): 601–3. doi:10.1111/j.1365-2621.1991.tb05338.x.

13. McGee, "The Science of Taste," 2011.

14. Parrish, Boles, Rust, Olson, "Dry and Wet Aging Effects," 1991, 601-3.

15. Nishimura, Toshihide, Rhue, Kato, "Components Contributing to Improvement," 1988, 2323–30.

16. Dashdorj, Amna, and Hwang, "Influence of taste components," 2015.

17. Savell JW. Dry-aging of beef, executive summary. National Cattlemen's Beef Association. 2008. http://www.beefresearch.org/cmdocs/beefresearch/dry%20Aging%20of%20beef.pdf [Ref list]

18. Warren, K E, and C L Kastner. 1992. "A Comparison of

Dry-Aged and Vacuum-Aged Beef Strip Loins." *Journal of Muscle Foods* 3 (52): 151–57. doi:10.1111/j.1745-4573.1992. tb00471.x.

19. Cook, Rob. 2018. "World Beef Consumption Per Capita (Ranking of Countries)." *Beef2Live*. January 24. http:// beef2live.com/story-world-beef-consumption-per-capita-ranking-countries-0-111634.

20. Dashdorj, Dashmaa, Vinay Kumar Tripathi, Cho Soohyun, and Inho Hwang. 2016. "Dry aging of beef; Review." *Journal of Animal Science and Technology* 58 (20). doi:10.1186/ s40781-016-0101-9.

21. Dashdorj, Tripathi, Soohyun, Hwang, "Dry aging of beef; Review," 2016.

22. "Argentina Beef." 2017. *Asado Argentina*. Accessed October 1. http://www.asadoargentina.com/argentina-beef/.

23. Steven Raichlen, and Matt Crowley. 2017. "Wet-Aging vs. Dry-Aging and How to Dry-Age Beef at Home." *Steven Raichlen's Barbecue! Bible*. Accessed December 4. https://barbecuebible.com/2016/03/15/ how-to-dry-age-beef-at-home/.

A great resource on barbecue and grilling in general, Raichlen's website also has wonderful information on how to safely and deliciously dry-age beef at home. This article draws from the expertise of Matt Crowley, the CEO of Chicago Steak Company.

24. López-Alt, J. Kenji. 2013. "The Food Lab's Complete Guide to Dry-Aging Beef at Home." *Serious Eats*. March. http://www.

seriouseats.com/2013/03/the-food-lab-complete-guide-to-dry-aging-beef-at-home.html.

Another indispensable resource on cooking meat (and just about everything else), J. Kenji López-Alt's writing for Serious Eats is a must for any lover of beef. And if *The Food Lab* isn't on your bookshelf yet, it needs to be.

25. See Appendix I for a list of the 25 restaurants selling authentic Kobe beef in the United States.

26. "Tajiri Issue." 2017. *Wikipedia Japan*. July 3. https://ja.wikipedia.org/wiki/%E3%80%8C%E7%94%B0%E5%B0%B-B%E3%80%8D%E5%8F%B7.

There can be some terminology confusion around Tajiri/Tajima/Kuroge Washu. Tajiri is the name of a specific sire born in the Hyogo Prefecture in 1939. Tajima cattle are those descended from Tajiri that also reside in Hyogo Prefecture—that's why some Kobe, which must come from Hyogo, sometimes is called "Kobe from Tajima cattle." The breed of Tajima-line cattle is Kuroge Washu. And to make things more confusing (or maybe to simplify them?!), 99.9 percent of Kuroge Washu breeding mothers are descended from Tajiri, meaning that Tajima and Kuroge Washu cattle are genetically the same (but for .1 percent). Tajima-line cattle are descendants of Tajiri, a sire born in 1939 in the Mikata District of the Hyogo Prefecture. For original Japanese-language sources, see the following websites: http://www.ojirokanko.com/ojiro_mura/html/ushi.html; http://cgi3.zwtk.or.jp

27. Heitzeberg, Joe. 2017. "Why Is Kobe Beef So Famous?"

Crowd Cow. November 7. https://www.crowdcow.com/blog/
why-is-kobe-beef-so-famous.

  Joe's learnings are drawn from conversations
  with Japanese beef producers, retailers, and marketing associations.

28. See Appendix I

29. Olmsted, Larry. 2016. "Kobe Beef in the U.S. Is Basically
    a Huge Sham." *Bon Appétit,* July 12, 2016. https://www.
    bonappetit.com/entertaining-style/trends-news/article/
    kobe-wagyu-steak-myths.
    Olmsted is the author of *Real Food/Fake Food.*

30. Heitzeberg, "Why Is Kobe Beef Famous," 2017.

31. Wagyu beef is known for having a low melting point, due in
    part to its high quantity of monounsaturated fats. See: "Kobe
    Beef." 2018. *Smart Kitchen.* Accessed January 29. https://www.
    smartkitchen.com/resources/kobe-beef.

32. Ari Notis. 2017. "The Nine Restaurants in America That Serve
    Real Kobe Beef." *Best Life.* August 21. http://bestlifeonline.
    com/kobe-beef-restaurants/.

33. "Inspection & Grading of Meat and Poultry: What
    Are the Differences?" 2014. *USDA Food Safety and
    Inspection Service.* USDA. June 3. https://www.
    fsis.usda.gov/wps/portal/fsis/topics/food-safe-
    ty-education/get-answers/food-safety-fact-sheets/
    production-and-inspection/inspection-and-grad-
    ing-of-meat-and-poultry-what-are-the-differences_/
    inspection-and-grading-differences.

"Utility, Cutter, and Canner grades are seldom, if ever, sold at retail but are used instead to make ground beef and processed products."

34. Ibid

35. Hale, Dan S, Kyla Goodson, and Jeffrey W Savell. 2013. "USDA Beef Quality and Grades." *Texas A&M AgriLife Research*. March 8. https://meat.tamu.edu/beefgrading/. The authors herein explain that beef carcass quality grading is based on "(1) degree of marbling and (2) degree of maturity."

36. Schatzker, *Steak: One Man's Search*, 2010.

37. Ibid

38. Addison, Bill. 2017. "The Best Restaurants in America 2017." *Eater*. November 8. https://www.eater.com/2017/11/8/16598768/best-restaurants-america-2017.

39. "Breeds of Livestock - Charolais Cattle." 2003. *Breeds of Livestock, Department of Animal Science*. Oklahoma State. February 23. http://www.ansi.okstate.edu/breeds/cattle/charolais/. Oklahoma State's livestock breeds repository is one of the best online sources for information on specific cattle breeds.

40. "Know Your Beef: Cattle Breeds That Make the Cut." 2015. *The Australian*. November 17. http://www.theaustralian.com.au/life/food-wine/

know-your-beef-cattle-breeds-that-make-the-cut/
news-story/ab4774f253e04d50b0749e94379ffe2a.

41. Ibid

42. "Grass-Fed Beef." *H. W. McElroy Ranch.* https://hwmcelroy-ranch.com/beef/.
This is said of the Maine-Anjou breed: "Fullblood Maine-Anjou's potentially have the best carcass offered in today's cattle industry. In different countries top Chefs are familiar with the term 'Anjou Beef.' The fullblood Maine-Anjou meat has one of the highest marbling scores and offers a high percentage of top end cuts compared to other breeds. With a considerable percentage of cutability (64% to 67%) and big rib eyes, these carcasses are in high demand."

43. "Breeds of Livestock - Maine-Anjou Cattle." 2003. *Breeds of Livestock, Department of Animal Science.* Oklahoma State. February 23. http://www.ansi.okstate.edu/breeds/cattle/maineanjou/

44. "Breeds of Livestock - Limousin Cattle." 2003. *Breeds of Livestock, Department of Animal Science.* Oklahoma State. February 23. http://www.ansi.okstate.edu/breeds/cattle/limousin/index.html/

45. *The Australian,* "Know Your Beef: Cattle Breeds," 2015.

46. *Breeds of Livestock, Department of Animal Sciences,* "Breeds of Livestock - Limousin," 2003.

47. Olmsted, Larry. 2016. "The Best (and Worst) of Times for Japanese Wagyu Beef in the United States." *Serious Eats.* July.

http://www.seriouseats.com/2016/07/fake-kobe-wagyu-beef-japanese-steak.html.

48. Ibid

49. Learned through Caroline Saunders's and Joe Heitzeberg's conversations with beef producers, processors, and companies in Japan.

50. Matthew. 2013. "Beef Myths Busted." *Tokyo Weekender.* May 16. http://www.tokyoweekender.com/2013/05/beef-myths-busted/.

51. Takafumi, Gotoh, and Sean-Tea Joo. 2016. "Characteristics and Health Benefit of Highly Marbled Wagyu and Hanwoo Beef." *Korean Journal for Food Science of Animal Resources*36 (6): 709–18. doi:10.5851/kosfa.2016.36.6.709.

The authors describe that fatty acid composition differs depending on breeds. They explain that highly marbled Wagyu has a higher proportion of monounsaturated fatty acid (MUFA) due to higher concentrations of oleic acid. They further explain that MUFAS "are heart-healthy dietary fat because they can lower low-density lipoprotein (LDL)-cholesterol while increasing high-density lipoprotein (HDL)-cholesterol."

CHAPTER 3: THE OTHER 88 PERCENT

1. McCarley, Brad. "Phone interview with Brad McCarley." Telephone interview by author. October 4, 2017.

All quotes hereafter attributed to Brad McCarley are direct quotes and have been checked by him for accuracy.

2. Forest, Susanna. "The Troubled History of Horse Meat in America." *The Atlantic*, June 8, 2017. Accessed September 3, 2017. https://www.theatlantic.com/technology/ archive/2017/06/horse-meat/529665/.

3. "Livestock and Poultry: World Markets and Trade." USDA Foreign Agricultural Service. October 12, 2017. Accessed November 17, 2017.
Our figure 12.3 million tons number is a projection of a 3 percent anticipated increase following continued herd expansion.

4. Severson, Kim. "Young Idols with Cleavers Rule the Stage." *New York Times*, July 7, 2009. Accessed September 4, 2017. http://www.nytimes.com/2009/07/08/dining/08butch. html?pagewanted=1&_r=2&partner=rss&emc=rss.

5. Ibid

6. Andrews, James. "Imports and Exports: The Global Beef Trade." Food Safety News. November 18, 2013. Accessed September 22, 2017. http://www.foodsafetynews.com/2013/11/ imports-and-exports-the-beef-trade/#.Wil6aLaZNsM.

7. Landers, Jackson. "A Dog-Eat-Dog World." *Slate*, April 19, 2013. Accessed November 18, 2017. http://www.slate. com/articles/health_and_science/science/2013/04/ what_is_in_pet_food_zoo_animals_sick_livestock_dogs_ and_cats_from_shelters.html.

8. Klinkenborg, Verlyn, and Andrea Modica. "Cow Parts." *Discover*, August 1, 2001. Accessed December 26, 2017. http:// discovermagazine.com/2001/aug/featcow.

9. Johnson, Brock. "Interview at Dahlia Lounge." In-person interview by author. August 2017.

Quote has been reviewed for accuracy by Chef Johnson and is reprinted, along with his full name, with permission.

10. Gerrard, Gene. "Vacio Steak or Bavette Steak." *The spruce* (web log), February 17, 2017. Accessed January 3, 2018. https://www.thespruce.com/vacio-steak-or-bavette-steak-2313863.

11. Ibid

12. Ramón II, Arturo. "Phone interview with Arturo Ramón II." Telephone interview by author. August 20, 2017.

All quotes hereafter attributed to Arturo are direct quotes and have been checked by him for accuracy.

In 2018, Arturo will be launching a restaurant near Driftwood, Texas, called Blanco River Meat Company, dedicated to whole-animal grilling. Worth a visit!

13. "Beginner's Guide to Yakiniku: How to Cook Wagyu Beef." Gurunavi, Inc. August 9, 2016. Accessed January 1, 2018. https://gurunavi.com/en/japanfoodie/2016/08/guide-to-yakiniku.html?__ngt__=TT0deaccb42005ac1e-4a594e3d3jaEFvxCSrPKodb7OAJm.

14. "Morcilla—Blood Sausage." Asado Argentina. 2010. Accessed December 15, 2017. http://www.asadoargentina.com/morcilla-blood-sausage/.

15. Rautureau, Thierry. "Interview at Loulay." In-person interview by author. September 1, 2017.

All quotes hereafter attributed to Thierry are direct quotes and have been reviewed for accuracy.

16. Recipe reprinted with permission from Chef Thierry Rautureau.

17. Haspel, Tamar. "Is Grass-Fed Beef Really Better for the Planet?" *The Washington Post*. February 23, 2015. Accessed July 10, 2017. https://www.washingtonpost.com/lifestyle/food/is-grass-fed-beef-really-better-for-you-the-animal-and-the-planet/2015/02/23/92733524-b6d1-11e4-9423-f3d0a1e-c335c_story.html?utm_term=.621a99c1d56b.

18. Hall, John B., and Susan Silver. "Nutrition and Feeding of the Cow-Calf Herd: Digestive System of the Cow." Virginia Cooperative Extension. May 1, 2009. Accessed October 3, 2017. http://pubs.ext.vt.edu/400/400-010/400-010.html.

To quote the authors, "The microflora in cattle rumens are fairly adaptable if grains or other energy sources are introduced slowly over time."

19. Schatzker, *Steak: One Man's Search*, 2010.

20. Uhlenkott, Paul. "Email correspondence with Paul Uhlenkott." Email interview by author. October 16, 2017.

Grain-finished beef producers at Cottonwood Ranch in Front Royal, Virginia, Paul and his wife Kristin opted to use a barley-and-hops-based brewer's mash supplement to finish their Black Angus herd instead of a cheaper, commercial corn/soy feed. Their choice, Paul explained, was largely about flavor. While corn/soy is more cost effective and puts on weight more quickly, brewer's mash puts on weight slightly slower but develops better intramuscular fat, or marbling, and a taste like molasses. They were aiming for flavor.

21. Lalman, David L., and Homer B. Sewell. "Rations for Growing and Finishing Beef Cattle." University of Missouri Extension. Accessed October 7, 2017. https://extension2. missouri.edu/G2066.

22. Uhlenkott, "Email correspondence with Paul Uhlenkott," 2017.

23. "Grazing Legumes and Bloat - Frequently Asked Questions." Alberta Agriculture and Forestry. April 28, 2003. Accessed November 8, 2017. http://www1.agric.gov.ab.ca/$department/ deptdocs.nsf/all/faq6752.

24. "Acidosis." 2015. *Beef Cattle Research Council*. August 13. http://www.beefresearch.ca/research-topic.cfm/ acidosis-63#acute.

25. Pollan, Michael. n.d. "So When Will Your Cow Meet Its End?" *PBS Frontline: Modern Meat*. https://www.pbs.org/ wgbh/pages/frontline/shows/meat/interviews/pollan.html.

26. Alberta Agriculture and Forestry, "Grazing Legumes and Bloat," 2003.

27. "Acidosis." 2015. *Beef Cattle Research Council*. August 13. http://www.beefresearch.ca/research-topic.cfm/acido-sis-63#acute. At Lil' Ponderosa Ranch near Carlisle, Pennsylvania, grass-finished beef farmer Uncle Bob explained that you need to keep legume levels below 30 to 40 percent in pastures to prevent bloat.

28. Holden, Ronald. 2018. "Grass-Fed Beef Loses Its Luster." *Forbes*, January 13, 2018. https://

www.forbes.com/sites/ronaldholden/2018/01/13/
grass-fed-beef-loses-its-luster/#270f380014dd.

29. Smith, Aaron. 2012. "Cash-Strapped farmers feed candy to cows." *CNN*. October 10. http://money.cnn.com/2012/10/10/news/economy/farmers-cows-candy-feed/index.html. Feeding candy to cattle is a common practice among some farmers during periods of drought, when corn prices skyrocket.

30. Saunders, Caroline, and Daniel Vaughn. 2017. Interview with Daniel Vaughn. Personal. All quotes are direct quotes from Daniel Vaughn, reviewed for accuracy and printed here with permission.

31. Saunders, Caroline, and Mely Martinez. 2017. Interview with Mely Martinez. Personal. All quotes are direct quotes from Mely Martinez, reviewed for accuracy and printed here with permission.

32. Ferdman, Roberto A. 2015. "I tried to figure out how many cows are in a single hamburger. It was really hard." *Washington Post Wonkblog*, August 5, 2015. https://www.washingtonpost.com/news/wonk/wp/2015/08/05/there-are-a-lot-more-cows-in-a-single-hamburger-than-you-realize/?utm_term=.cod2e3d38bb0.

33. Cordain, Loren. 2014. "Tongue: A Hunter Gatherer Delicacy." *The Paleo Diet*. November 3. https://thepaleodiet.com/tongue-hunter-gatherer-delicacy/.

34. Recipe is reprinted with permission from Mely Martinez.

35. Saunders, Caroline, and Meathead Goldwyn. 2017. Interview

with Meathead Goldwyn. Personal. Quote has been reviewed for accuracy and is printed here with permission from Meathead Goldwyn.

36. Saunders, Caroline, and Brad McCarley. 2017. Interview with Brad McCarley. Personal. All quotes are directly quoted from Brad McCarley, have been reviewed for accuracy, and are printed here with permission.

37. Severson, Kim. 2009. "Young Idols with Cleavers Rule the Stage." *New York Times.* July 7. http://www.nytimes.com/2009/07/08/dining/08butch.html?pagewanted=all.

38. Ibid

39. Ibid

40. Ibid

CHAPTER 4: A SECOND CHANCE FOR SMALL FARMS

1. Heitzeberg, Joe, Ethan Lowry, Caroline Saunders, and Becky Harlow Weed. 2017. Interview at Harlow Cattle Co. Personal.

   All quotes herein printed are direct quotes from Becky, and have been reviewed by her for accuracy. They are printed here and her full name is used with permission.

2. Olmstead, Gracy. 2016. "Down on the Farm." *The National Review,* August 15, 2016. http://www.nationalreview.com/article/438983/small-farms-big-business-family-farms-struggle-against-industrial-agriculture.

3. "Background." 2017. *United States Department of Agriculture*

Economic Research Service. January 23. https://www.ers.usda.
gov/topics/animal-products/cattle-beef/background.aspx.

4. Ibid

5. "Beef Industry Statistics." 2017. *National Cattlemen's Beef
Association.* Accessed August 20. http://www.beefusa.org/
beefindustrystatistics.aspx.

6. Mathews, Kenneth, and William D McBride. 2011. "The
Diverse Structure and Organization of U.S. Beef Cow-Calf
Farms." *United States Department of Agriculture Economic
Research Service.* March. https://www.ers.usda.gov/
publications/pub-details/?pubid=44532.

7. Pollan, "So When Will Your Cow Meet Its End?"
PBS Frontline.

8. "Slaughter-Dressing of Livestock." 2017. *Texas A&M
University.* Accessed August 10. https://meat.tamu.edu/
ansc-307-honors/slaughter-livestock/.

9. Reprinted with permission from Uncle Bob and Kate Boyce.

CHAPTER 5: UNSUNG ENVIRONMENTAL HEROES

1. Saunders, Caroline, George Lake, and Christy Lake. 2017.
Interview at Thistle Creek Farm. Personal.

  All quotes herein are directly quoted from George and
Christy, have been reviewed by them for accuracy, and are
printed here, including full names, with permission.

2. McCulley, Rebecca. 2017. "[Transcript] Grasslands and
Carbon: Processes and Trends." *United States Department of
Agriculture.* Accessed November 5. https://www.fs.usda.gov/

ccrc/sites/default/files/carboncourse/transcripts/4.McCulley.
pdf.; Fynn, A. J., P. Alvarez, J. R. Brown, M. R. George, C.
Kustin, E. A. Laca, J. T. Oldfield, T. Schohr, C. L. Neely, and
C. P. Wong. n.d. Chapter IV: Soil Carbon Sequestration
In United States Rangelands. Chapter IV: Soil Carbon
Sequestration In United States Rangelands. http://www.fao.
org/docrep/013/i1880e/i1880e03.pdf.

3. Ibid
4. Ibid
5. Schwartz, Judith D. 2014. "Soil as carbon storehouse:
   New weapon in climate fight?" *Yale Environment
   360*. March 4. https://e360.yale.edu/features/
   soil_as_carbon_storehouse_new_weapon_in_climate_fight.

   Haspel, Tamar. "Is Grass-Fed Beef Really Better for the
   Planet?" *The Washington Post*. February 23, 2015. Accessed
   July 10, 2017. https://www.washingtonpost.com/lifestyle/
   food/is-grass-fed-beef-really-better-for-you-the-animal-and-
   the-planet/2015/02/23/92733524-b6d1-11e4-9423-f3d0a1e-
   c335c_story.html?utm_term=.621a99c1d56b.

6. Lela, Nargi. 2017. "New Study Shows Organic Farming
   Traps Carbon in Soil to Combat Climate Change." *Civil
   Eats*, September 11, 2017. https://civileats.com/2017/09/11/
   new-study-shows-organic-farming-traps-carbon-in-soil-to-
   combat-climate-change/.

   Email correspondence with The Organic Center's
   (TOC's) Director of Science Programs, Jessica Shade, is
   cited herein.

7. Saunders, Caroline, and Fred Colvin. 2017. Interview at Colvin Ranch. Personal.

8. Poem is reprinted with permission from George Lake.

CHAPTER 6: A FLIGHT OF BEEF

1. Saunders, Caroline, and Robert Schutt. 2017. Phone interview with Robert Schutt. Personal.

2. Saunders, Caroline, and Kurt Beecher Dammeier. 2017. Steak tasting at The Butcher's Table. Personal.

3. You can see Kurt's own rating system for his beef on his website, at the following url: http://www.mishimareserve.com/#our-story/

4. Miller, "Dark, Firm and Dry Beef," *Beef Research*, 2017, 1–3.

5. Saunders and Meyers, Interview at Sweet Grass Farm, 2017.

6. Schulz, Eric. 2017. "An Introduction to the Maillard Reaction: The Science of Browning, Aroma, and Flavor." *Serious Eats*. April. http://www.seriouseats.com/2017/04/what-is-maillard-reaction-cooking-science.html.

WWW.CROWDCOW.COM

Questions? Write to: craftbeefbook@crowdcow.com